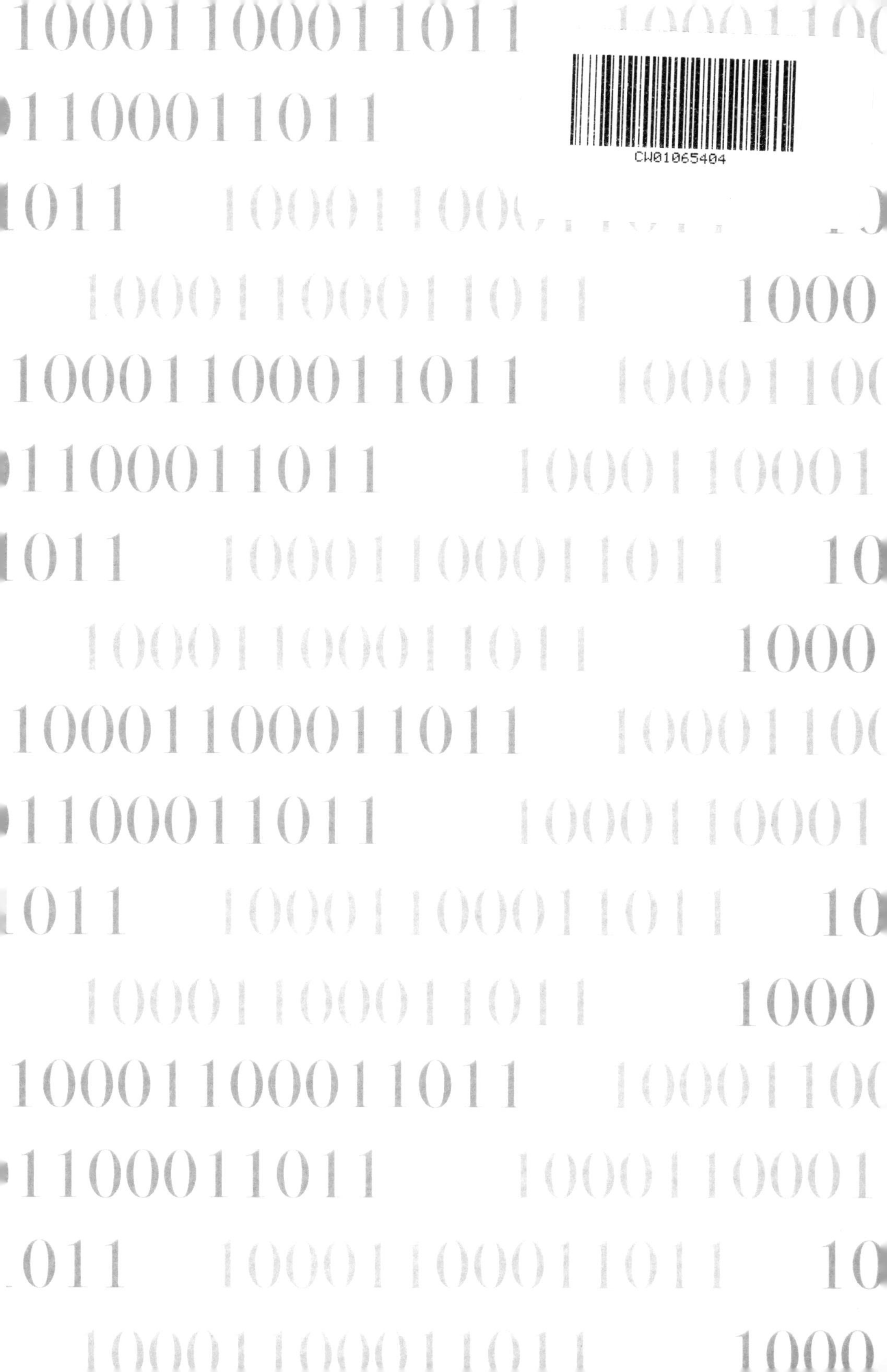

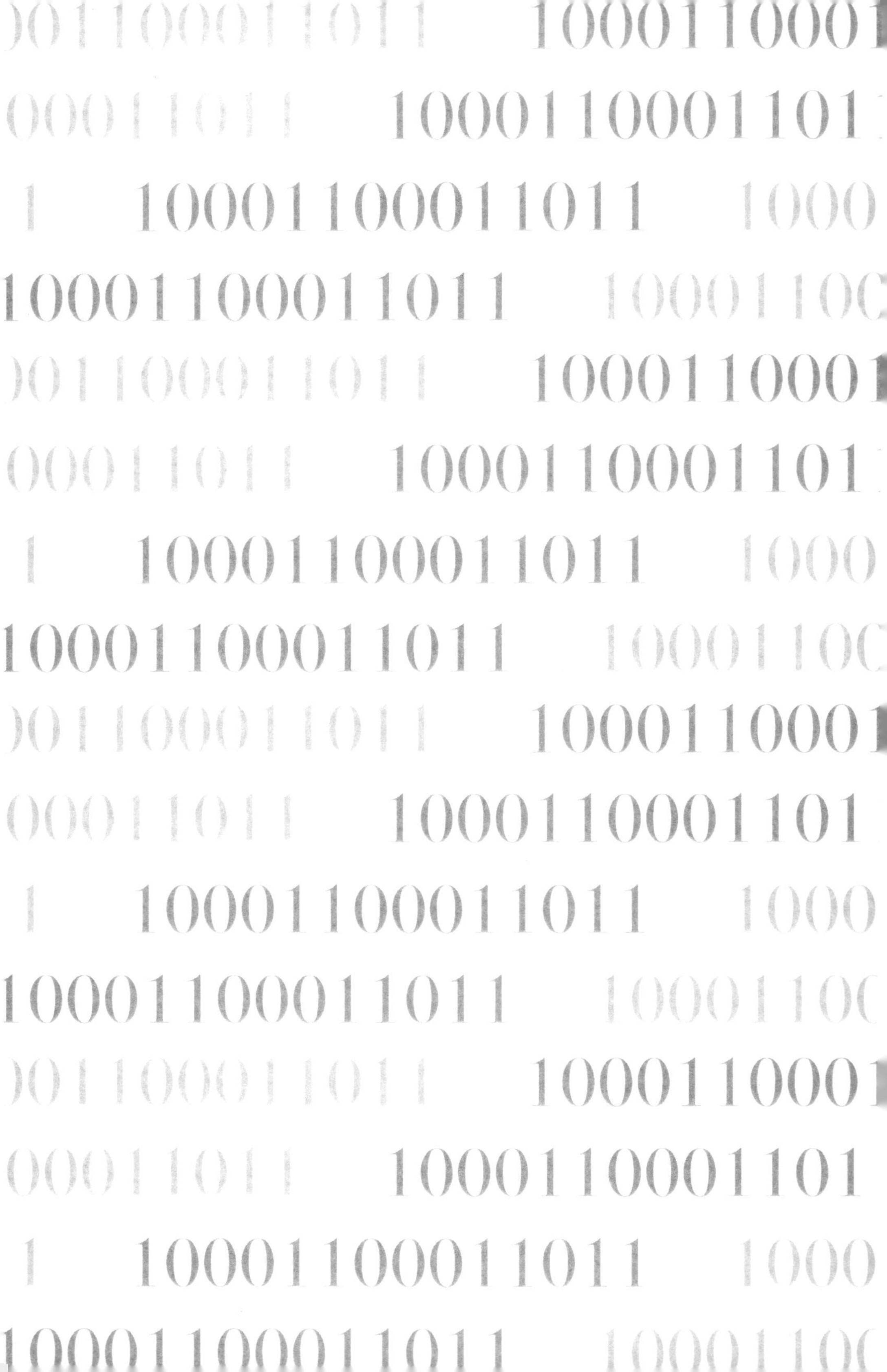

PRAISE FOR Decoding AI in Financial Services

"Every aspect of The Global Financial Services industry is transforming at machine speed as incumbents and disrupters look to capture the hearts, minds, and wallets of this next generation. Clara Durodié's book is a must read and essential guide for all, including boards, venture capitalists, operators, and academics looking to understand how AI is transforming Financial Services. She succinctly captures the current state, clarifies how it is all reshaping, and concludes with how we deal with what's next."
–Ron Suber, Chairman of Credible and Fintech board adviser and investor, San Francisco, CA, United States

"A comprehensive analysis of AI, its origins, uses, and application, both current and potential, for the financial services sector. This provides references for businesses, is thought-provoking, and is a useful guide for firms."
–Liz Field, CEO of Personal Investment Management & Financial Advice Association, London

"This is a remarkable book and documentary on artificial intelligence. It is comprehensive but very understandable for either the person who has little or no understanding or indeed for the expert. It is a must have. I come to the subject with little depth of knowledge or expertise. I leave with both, but also, an understanding of the history, progress, and future for AI. The book covers the technology, its application, as well as the ethics. It describes in very clear terms how business and society progressively will be unrecognisable in the next decade and beyond. The application of artificial intelligence will dominate every aspect of human endeavour and change our lives in a way as significant as the Industrial Revolution. The book addresses all these issues with ease and great comprehension. It also addresses the pitfalls but overwhelmingly the remarkable benefits and improvements it will bring, in particular to the Financial Sector. It will be essential reading for any business leader or Board Member."
–David Slessar OBE, Chairman of the Australian British Chamber of Commerce, Sydney, Australia

"Clara Durodié does a masterful job of explaining AI in simple terms and then lays out all the areas within financial services that will be transformed and disrupted by AI in the coming years and decades. I find her holistic, human value-centric view to be very compelling and timely as we go about understanding, applying, and scaling arguably the most powerful human innovation in history. Finally, her views on the leadership and organizational challenges associated with the

successful and responsible adoption of AI for maximizing societal benefit and minimizing harm is both very insightful and timely."
–Manoj Saxena, Chairman of Cognitive Scale and the first GM of IBM Watson, Austin, TX, United States

"I started reading this book with trepidation. There is so much hype around AI, so many fears, so much money spent, so little understanding of what it is and how to best approach it. So, how could one book do it justice and how could it take a neophyte from nothing to an understanding of what questions a Board should be asking? Well it does! Clara Durodié has written a book that demystifies the jargon, while comprehensively covering the subject from both a technical level to strategic issues. Unlike most practitioners, Clara balances the practical application of AI with a good understanding of the ethics involved, bringing into stark reality what can happen if we go AI mad without considering the consequences. There is a wealth of information and reference to further reading material for everyone from the new Chief Digital Officer to all the C-suite and the Board. I particularly like the plentiful use of current examples around the world of who is utilizing the different forms of AI and in what applications."
–John Meinhold, Independent Advisor and Financial Institutions Board Member, Kuala Lumpur, Malaysia

"For boards, the format is information rich, sharp, complete, and structured to help any director or practitioner get to the root of any contemporary AI challenge. The book cuts through the clouds of hype and misinformation surrounding the transformative powers of Artificial intelligence. Covering common AI terms, definitions, and capabilities essential to analyse or strategise this book provides clear questions and targeted frameworks to run, manage, partner, or acquire AI sector companies. Importantly, the future ecosystems evolving and the ethics that should underpin them are covered extensively in a digestible clarity I haven't seen before. This is an outstanding reference and extraordinary resource that will be a landmark publication. Pure value. Pure actionable insight."
–Colin Bennett, Head of Digital Distribution (Global), GAM Investments, London

"As AI becomes increasingly more prevalent within the financial services sector, *Decoding AI in Financial Services* is a must-read for companies that want to stay ahead of the curve. Clara Durodié is able to make sense of the most complicated of concepts, making this book an essential tool for those who are new to the AI space and those who are also driving the financial service industry's AI solutions. **–Donna M. Parisi, Derivatives Partner and Global Head of Financial Services & FinTech, Shearman & Sterling LLP, New York**

"Accessible and easy to read, it explains in just enough detail how AI is transforming the financial services sector. Highly recommended for senior executives."
–Ardi Kolah, Executive Fellow, Henley Business School and Privacy Consultant; Founding Editor-in-Chief, Journal of Data Protection & Privacy, London

"This is a book I would've liked to have written about artificial intelligence. I have seen the need for a number of years now to take the often mystifying and misconceived subject of artificial intelligence and put together a comprehensive and comprehensible form for the senior businessperson who seeks to be able to deal effectively with this disruptive set of technologies. With *Decoding AI in Financial Services*, Clara Durodié has condensed fifty years of progress, and decades of futurecasting, into a single resource for the board and for executive management, the go-to reference on what to do about AI."
–David Shrier, University of Oxford & MIT Futurist, Oxford Fintech programme director, Boston, MA, United States

"In this book, Clara Durodié bridges the world of research and business in a practical way that I found very useful. I'm sure this book will be widely read in boardrooms across many sectors."
–David Birch, best selling author, commentator on digital financial services, author of Before Babylon, Beyond Bitcoin and Identity is the New Money, London

"Clara Durodié's book is one of the most thoroughly-researched and well-written books on AI out there, written specifically for the business reader at a time when AI is changing the world, and more specifically the service delivery models of most financial organisations and institutions. The book manages to stay clear of the noise and hype, and deliver a structured approach for understanding the risks, complexities, ethics, and opportunities of automating the human intellect in the 21st century. As such, it is highly recommended reading for anyone who wishes to understand, and successfully implement, AI in financial services."
– Dr George Zarkadakis, Digital Lead, Willis Towers Watson, science writer with a PhD in Artificial Intelligence and author of In Our Own Image, London

"Clara Durodié demystifies AI by outlining its value and applicability as a tool to unleash the power of data. It thoroughly examines the technical, ethical, and business implications of delivering AI at scale which will set the stage for the data driven future of financial services." **– Mike H. Blalock, General Manager, Financial Services Industry Vertical, Intel Corporation, Boston, MA, United States**

Decoding AI in Financial Services

Business implications for Boards and Professionals

To Daniel

Enjoy reading the book!

First Edition

Clara Durodié

First published in Great Britain in 2019.

Hardback: ISBN 978-1-905912-66-7

Printed by Blissetts, Roslin Road, Acton, London, W3 8DH www. blissetts.com

Blissetts policy is to use paper that is natural, renewable, and recyclable produced from wood grown in sustainable forests. The logging and manufacturing processes are expected to conform to the environment regulations of the country of origin.

Cognitive Finance Press, London.

Cognitive Finance Press is committed to contributing to global reforestation. Each copy sold of *Decoding AI in Financial Services* pays for planting and caring for a tree until it is able to sustain itself.
Cognitive Finance Press publications are available for special promotions and premiums. For details contact: info@cognitivefinance.ai

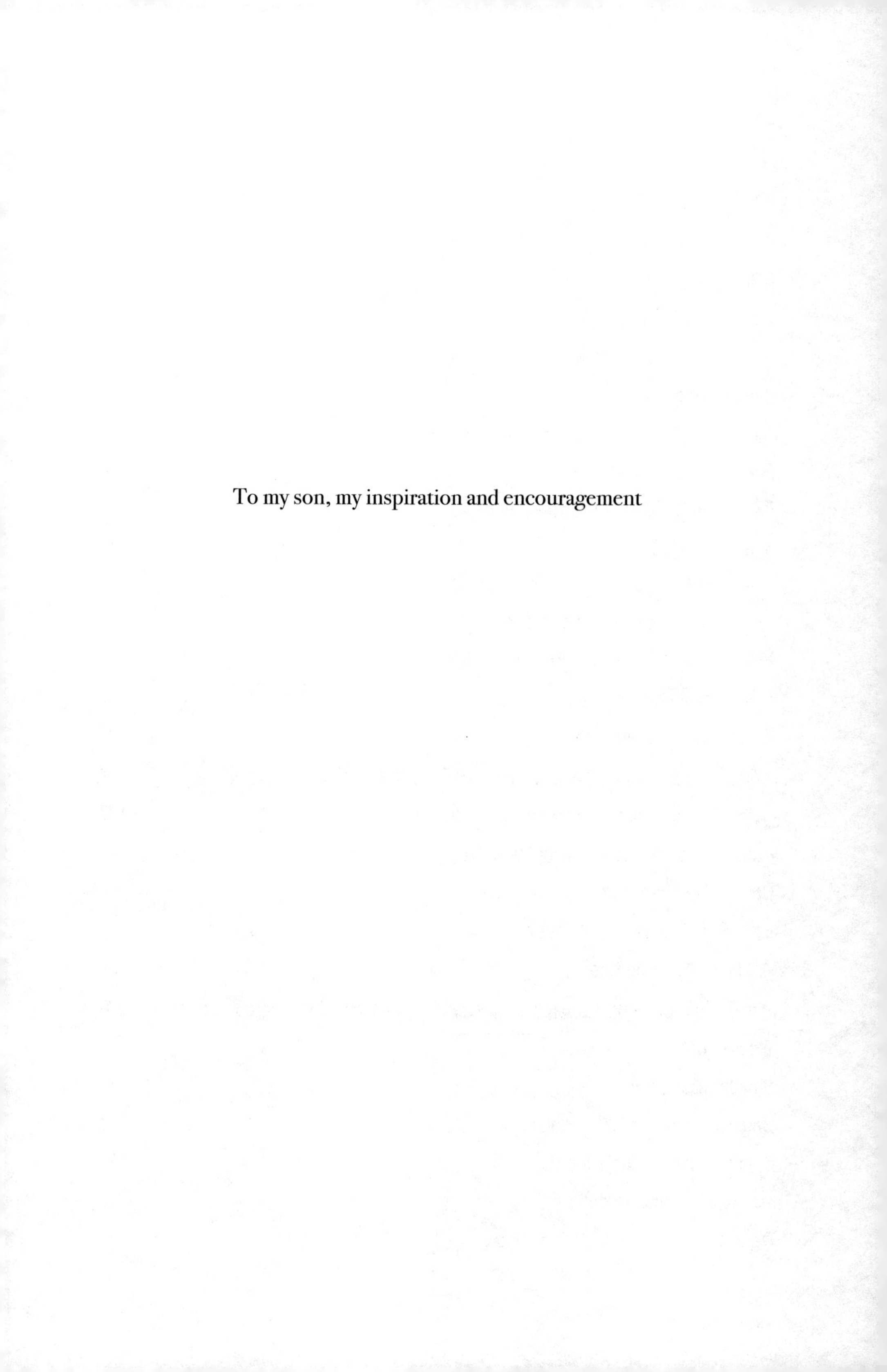

To my son, my inspiration and encouragement

Introduction

It was the end of a long day, and I felt tired. I took a quick look at the small group of board directors who were still talking and pointing at diagrams that I had prepared for this meeting. They looked tired, too. It was the second day of an AI strategy work group at a reputable bank's board meeting.

Unlike yesterday, the chairman had been quiet during today's work group. *Quiet minds think loud,* I thought to myself, as I saw him walking towards me.

"It was a great work session. You need to put all of this in a book because it is a lot to take in," he said, as his opening line to what turned out to be a two-hour conversation on new revenue models in an AI-driven business, the future . . . and the genesis of this book.

WHY I WROTE THIS BOOK

Writing a book has never been on my to-do list. However, having worked with a large number of boards, I have come to appreciate that a typical board director in financial services is bombarded by an ever-increasing volume of information that is large, overwhelming, and in many cases paralysing and intimidating for non-technologists, which form the majority of financial boards.

I have written this book to empower boards, decision makers, and professionals to make sense of this vast volume of information and put it in the context of our industry. The book is a primer that identifies a broad range of topics that boards need to focus on. The book's content is aligned with the feedback that I received from board directors and industry professionals who identified a need for a structured, impartial, and fact-based reference of what they need to know. Therefore, the book remains impartial, provides baseline information, directs to further reading and provides a general blueprint for critical thinking. The book doesn't get deep into technical details or evaluate AI vendors and business strategies. This book is an introduction to a journey that our industry has already embarked on, and includes some lessons learned so far. It aims to empower you with a roadmap and enough information, so you can start your own AI journey.

As a former industry management executive trained as a portfolio manager and who also worked in business management roles, this is the book I would like to have had as reference material to help me navigate a sea of AI confusion, hype, and new risks that are upon our industry. My engagement with technology harkens back to 1998, when I worked with early forms of barcode technology, inventory optimisation and personal credit rating. In 2003, when I needed client data analytics, and when no one could provide it, because no one had ever done it before, I did it myself. In 2011, I suggested in a management committee meeting that the organisation should invest in intelligent automation rather than traditional software for back office automation and client reporting. It was forward thinking and correct at a time when the industry used AI primarily in trading, portfolio management and fraud detection. I resigned in 2014, to establish Cognitive Finance Group and to research trust with technology as well as the intersection of AI, neuroscience, and wealth management. In 2015, I became a board adviser specialising in artificial intelligence (AI) in financial services with Cognitive Finance Group. In my current role, I work with the boards of leading financial services institutions on defining business transformation with an AI enterprise-wide strategy, risk management and cybersecurity. I also assist with AI vendor selection and work on select data-science projects which give me direct access to implementation challenges. In addition to this direct experience, I am a globally recognised thought leader in applied AI in financial services, keynote speaker at leading international financial services conferences and have received industry accolades for my work. I also advise investment funds and PE houses on their investments in AI.

I mentor AI start-ups on product development, ethics, funding and growth strategy. I have a masters degree in financial management from Oxford University's business school, where I teach a class in their Artificial Intelligence Program. I have advised organisations and executives from all continents, including the World Economic Forum, the Japanese government's special AI commission, the UK's Government All Parliamentary AI committee, a host of leading western and Asian banks, the Institute of Directors or equivalent in North America and the Far East, and chairing the AI committee responsible for drafting a paper on good practice for a UK's leading Non-Executive Directors network. I am fluent in the world of business strategy with AI, and because of my expertise I also know that ethics is an important piece of the AI puzzle. Ethical AI can also deliver some of the ESG agenda. Armed with this experience, I believe that I have a unique perspective to share. This perspective is especially important, given that financial institutions need to embrace AI with confidence, knowledge, and trust in order to remain relevant and foster growth and financial stability for our society. After two years of research and writing, interviewing 180 executives, about 15,000 bibliography entries, just under two months spent in the University of Oxford's libraries completing in-depth academic research to validate my assumptions, and addressing more than 300 industry conferences, I am able to present you with this book. Writing this book has taken longer than planned, partly because I did the research alone, without the support of junior analysts or researchers, but critically, because as the AI space is evolving so quickly, I had to cover more ground to ensure a thorough review and robust trend analysis. I have made a promise to you, my reader, that I will write it in simple terms and for this reason the book went through several iterations. My father has always reminded me that if I can explain complex concepts in simple terms, then that is the measure of how profoundly I understand the concepts. Simple is not always that *simple*.

AI is diverse and complex. AI in business transformation is even more complex, since it affects people and businesses. AI in financial services, a regulated industry, introduces even further complexity. It has real-world ramifications. For more than 60 years, the AI community has not agreed on a definition of AI. Writing a book that has a reductionist approach to the complexity of AI in business will need to forego much of the detail and this will inevitably frustrate the purists. So, for the purists out there–or those who are prone to unduly complicating things–I have a message for you: Hold it! Boards and decision makers in financial services are facing one of the most challenging tasks of their careers, namely, to transform their organisations into profitable technology businesses. As such, they first need the AI complexity and its business implications distilled to the simplest form. There is a clear trend that the ultimate responsibility for the correct use of AI rests with the board, which makes AI a board issue rather than just an IT issue. Therefore, the board and C-suite need to have a firm grasp of the core AI concepts, so that they can be confident to make their own informed decisions, thereby escaping the current paralysing confusion and fear of what these AI technologies mean. This books aims to meet this need, and presents an impartial reference.

WHO SHOULD READ THIS BOOK

This book will be of interest to anyone who wishes to quickly and competently know what lies ahead for this industry. This book takes you through current industry thinking and applications, so that you can be best prepared to comprehend important AI issues that are imperative to your business. This book is deliberately written in plain English. Financial services is a coded industry, with a range of complicated terminology that only professionals are able to decipher. The same is true of computer science. In order to open the conversation and allow my readers fair access to its knowledge, I have removed the jargon and simplified concepts where possible. Board directors need to know enough about AI to keep their organisation safe on the road to AI adoption - they do not need to learn how to code. AI is not an option–it is a necessary business tool to keep organisations competitive.

This book is written anticipating your time availability:

- *Do you have very little time ahead of a meeting?* The Key Points to Remember at the end of each chapter might provide a useful summary.
- *Do you have a meeting to discuss AI and do not know what questions to ask?* At the end of each chapter, there are a set of questions to jumpstart most AI-related conversations and provide you with a useful prompt for areas to explore further. I recommend that you carefully evaluate the answers that you receive to these questions.
- *Do you have an interest in a particular area?* The Table of Contents is your roadmap to find what you are interested in and to drill down into that chapter.
- *Are you ready to bring AI to your organisation?* I would recommend reading the book in full as it will empower you with the knowledge you need to start your own AI journey with confidence.
- *For time-management purposes:* reading the whole book should take you about 15 to 20 hours. I recommend that you keep the book as a reference, so you can always return it to refresh your memory or reread concepts that you may not have fully understood during your first reading.

This book also provides valuable insights for a range of professionals including:

- Financial services practitioners, irrespective of their level of seniority
- Government and ministry of finance officials, central bankers, regulators, and trade associations.
- Students and academic researchers to give them an insight into the challenges that this industry has to overcome in order to bring to the market trustworthy technology.
- Personnel and recruitment specialists, especially board recruitment specialists who systematically fail to appreciate how essential it is to advise their clients to include

experts who are fluent in applied AI and business strategy and have financial sector experience, on their corporate boards.

- Legal advisers, as their work is critical to their financial services clients.
- Marketing specialists who are serving financial services clients.
- AI start-ups looking to target financial services to understand the challenges this industry has and how their competitors are faring.
- Investment professionals from venture capital investors (to understand the specific challenges financial services faces in adopting AI solutions) to traditional fund managers. Having advised a range of VC funds and PE houses looking to invest in AI start-ups, I have come to appreciate that they would benefit from having industry-specific knowledge of how AI is adopted.
- Other industry professionals like telecoms, who would like to understand how AI can benefit their organisations.

A BUSINESS CASE FOR ETHICAL AI

Our industry is being drowned in data that is neither being used correctly nor benefiting fully our organisations and customers. This book aims to showcase how to use data and do so ethically. It also intends to clarify how you can use technology as a trustworthy platform, so that you can confidently say that you deploy "ethical AI for business growth and profitability."[1]

Joined up, strategic ethical deployment of AI for profitability and growth is the next frontier for successful organisations in our industry. The clients I have been working with are already realising the benefits of this strategic approach. For instance, the current technology allows banks to predict when a customer is at high risk of default. Instead of letting them default and then start a chain of convoluted and expensive debt collection procedures accompanied by threats to pursue them in court while damaging their credit rating, a bank can write to those clients and say something like "if you are experiencing difficulties with the repayment, we are here to help and offer you a two-month holiday payment until you get back on your feet. It is possible. It can be done, and it has a long-term positive outcome.

In London, Cognitive Finance Group conducted ad-hoc research of 150 randomly selected people travelling on the London underground. We tried to ascertain if these people would consider leaving their bank if they were offered a payment holiday in a default on loan scenario. Ninety-six percent said that they would not leave that bank. Ethical AI builds loyalty and trust. In the UK, in 2019 the FCA updated the Senior Managers and Certification Regime with new responsibilities that might be addressed with deployment of ethical AI.

WHAT YOU WILL LEARN

This book aims to provide clarity and structure. It is a non-technical introduction to AI in business and it is written in an accessible style that minimises technical jargon and deliberately does not include mathematical equations. This book deliberately stays away from providing solutions and resolutions. They can only be general and counterproductive. Each organisation is different. What works for one organisation, doesn't necessarily work for others. For the same reason, it was a deliberate decision to share a list of vendors but not to evaluate them or explicitly share solutions from my hands-on experience.

Nonetheless, the questions that boards can ask are drawn from my hands-on experience, yet I chose not to want provide the answers, because each organisation is different and generalising solutions is not an approach that has proven constructive. I could only provide answers to these questions if I am consulted with respect to an organisation's particular needs.

Some of the key points you will learn from this book include:

- What AI does and does not mean, as well as key terminology
- Types of machine learning, their challenges and applications
- How to source and collect datasets, including some of the top open source datasets
- Which principles to follow in order to become a successful data-driven business
- Why to invest in AI and a recommended due diligence framework to follow
- How AI can help fintech with data storage, mobile services, fraud detection, and cybersecurity
- Why you should incorporate AI into your business strategy and where to start
- The difference between real AI products and pseudo-AI products
- Why it is essential to have a framework for understanding the particularities of due diligence in AI transactions
- The five important facets of AI governance and the ethical guidelines to follow
- How algorithmic bias occurs and how it can be corrected

- Why privacy will be one of the leading technology themes and why boards and executive teams should focus on how the AI tools they are using are safeguarding data
- Why boards should ask the executive for a 10-year roadmap for AI adoption and should have oversight over this strategy
- The key risks and legal issues for boards to monitor
- How AI can enhance board productivity by saving time, making meetings more effective, and providing real-time data for decision making
- Key benefits of AI like scalability, personalisation, and speed–and how they are already being used in financial services
- How to handle internal resistance to change
- Why the absence of a coherent AI strategy is a business risk and technology must follow the business strategy–not the other way around
- A snapshot of the current trends and challenges of AI in investment banking, retail banking, and asset & wealth management
- Use cases in customer service, portfolio management, sales and marketing, personnel, legal, risk management and compliance, fraud prevention, finance, and IT

HOW THIS BOOK IS ORGANISED

Chapter 1 of this book starts with addressing the question "what is intelligence?" from theological and philosophical approaches. Why? Humanist disciplines inform us to never lose track of one of the core tenets of this book: Technology must empower humanity, and if we do not know what humanity stands for and the history of its beliefs–in particular, that one about the supreme intelligence–we will never be able to understand and design technology that serves humanity and furthers an organisation's growth. This chapter also discusses seemingly unconnected concepts like emotions and consciousness, which are actually regarded by some as the ultimate test for conferring AI tools legal personhood–a slippery slope for the world not only in our industry but across society, because it provides a legal shield for bad actors. The chapter also covers a summary of the philosophical reflections on technology, free will, agency, and design.

Chapter 2 discusses what AI is, the basic concepts, and the infrastructure needed to deploy AI in the enterprise. It also outlines the taxonomy of AI technologies and provides in-depth explanations of how machine learning and deep learning operate, as well as hybrid models.

Chapter 3 focuses on the preconditions of AI, with information on data, computing power, cloud, AI-as-a-Service, and AI-enabling infrastructure design and thinking.

Chapter 4 discusses the current state of AI and presents maturity cycles, due diligence when investing in AI, and geopolitical considerations as AI defines global geopolitics.

Chapter 5 is the longest in the book because it covers the current risks and ethical challenges including privacy, bias, deep fakes, authentication, economic and humanitarian issues, as well as antitrust and competition law, AI safety, and Artificial General Intelligence. This chapter is particularly essential for boards and decision makers, so they are equipped with the right knowledge of the deep ethical concerns and risks surrounding AI. When a few C-suite members asked me why they should concern themselves about ethics in AI design, I realised that the leadership would not be able to make the right decisions unless they understood how fundamental ethics is in the AI-driven enterprise.

Chapter 6 empowers board directors with specific knowledge about the oversight of AI, not only the corporate governance of AI but also using AI to increase the efficiency of boards.

Chapter 7 focuses on the strategic adoption of AI in the enterprise, addressing core concerns like how to build an AI business strategy, where to start, and how to re-skill the workforce.

Chapter 8 highlights broad trends in various sectors as to how AI functions in financial services in wealth management, asset management, retail banking, financial markets, payments, and insurance, as well as covering specific uses cases like mutuals, islamic finance, and also how central banks and regulators are benefitting from AI.

Chapter 9 zooms in on the process across the enterprise, looking at areas like portfolio management, legal, compliance, customer service, marketing, finance, and IT to showcase a wide variety of use cases and help you gain a practical understanding of how this technology is changing each business process.

The final chapter, *Chapter 10*, is a compilation of final remarks on how the future is being shaped by AI and highlights a range of issues that we need to be aware of.

At the end of each chapter, you will find two sections:

1. **Examples of Questions for Boards to Ask.** These questions are by no means exhaustive, they highlight some of the most important points for boards to consider

based on the material within each chapter. When your organisation is beginning the process of implementing AI, these questions will help you to focus in on some of the anchor points to investigate when making AI investments and deployment decisions.

2. **Key Points to Remember** is the final section that summarises the material within each chapter and provides fundamental ideas for you to remember.

AUTHOR'S NOTE

Throughout the book we have decided to use data as singular, in order to simplify the reading but also to reflect the journalists' preference, rather than adhering strictly to Latin grammar rules, which advocate for datum (singular) and data (plural).

The truth can be construed as contentious when it is inconvenient. My work including the vast amount of research provide a solid foundation for a truthful account of the topics that I cover. I believe that decision-makers need "radical honesty"[2] when confusion and hype distort reality. This book provides honest and unvarnished opinions about the current state of AI adoption; it covers candidly some executives' ineptitude and others' genius, and the damaging creativity found in some vendors's marketing teams up-selling their technologies. This is never meant to disparage, but to highlight biases, practices, and fears that need to be addressed correctly, so they lift rather than lower expectations. I expect these communities to react with criticism. Why did I decide to criticise them? This book is primarily written for decision-makers who sign off the digital innovation budgets which pay consultants, cloud providers and technology vendors.

With 75 percent of digital innovations projects going to waste or not delivering what they promise (according to McKinsey's 2019 numbers), only 25 percent of organisations having an enterprise-wide AI strategy (IDC 2019 report) and 80 percent of banks leadership lacking any technology background (according to Cognitive Finance Group's 2018 research), boards would benefit from an easy to read, impartial and fact-based briefing so they can make their own business decisions, understand what they are committing to and what they can realistically expect to achieve. This impartiality or professional integrity is that much more important to regulators and boards of smaller organisations which do not have the vast resources and budgets that global organisations enjoy.

Now, it is time to begin our journey.

Let's discover how AI reshapes our industry, what opportunities lie ahead of us, and how to use ethical AI to maintain a profitable business across generations. At the time of writing this introduction, my country is shaken by deep political unrest as we approach our exit from the European Union family. On this backdrop, I found Winston Churchill's words suitable to also describe what AI now means to the financial services industry:

This is not the end. It is not even the beginning of the end.

But it is, perhaps, the end of the beginning.

CHAPTER ONE

What Is Intelligence?

In a book that deals with specifics, I believe it is important to take a step back and occasionally deal with the general concepts. Too often the bigger picture can be forgotten in investigation of the minute detail. Generally speaking, this is a reflection of a future-proofed mindset that can help identify links between disciplines surrounding science and humanities. This complex dynamic between two seemingly separate magisteria lies at the centre of creating a society in which artificial intelligence can coexist with humanity and it also impacts and shapes financial services. In fact, the understanding of science alongside a study of humanities can provide the ideal platform from which to best implement artificial intelligence (AI) systems in a specific context like financial services. These three disciplines can help enrich one another in a way that provides the most unique angles of perception.

Fundamentally, the study of AI needs to be placed within the context of history and the study of people. It is crucial to understand the impact that groundbreaking inventions have had on the past, to better understand the future. Especially, as many argue, because AI is the most disruptive technology to date, with its capabilities to both do well and cause harm. In this context, we can look towards some of the writings that argue that AI brings to light three fundamental questions to consider as the AI community is pushing hard to reach Artificial General Intelligence:

1. Is Artificial General Intelligence as the omniscient entity the ultimate goal for the AI field? Should it be?
2. What is a beneficial transition beyond humanity?
3. How do we get there from here without destroying ourselves?[3]

Some argue that humans are in a constant state of development towards an inevitable extinction where they are inherently selfish and immoral. I would argue that, of the multiple solutions presented for the second question, cognitive enhancement seems to be the most powerful transition beyond humanity, and bioengineering companies (discussed in Chapter 10) are the predecessors of that. However, for the sake of this section, I will focus on the third question and the historical lessons from which we can learn. It seems clear that at most notable turning points in Western Europe, there comes with it an extremist reaction in both an academic and political sphere. Often, the needs of wider society aren't taken into account, such as the devastating impacts of the Industrial Revolution on our climate. I, and I think my generation, would be keen to see an approach to AI that encourages legal foresight and a mindset for the longevity of development rather than the short-term gains. Again, I would argue that this outlook can only be encouraged by those who can take a step back from the technical details and understand the larger impact of such positives.

FEAR AND A CLASH OF MINDSETS

John O'Leary and John Kingston are professors at the Artificial Intelligence Applications Institute at the University of Edinburgh. In their insightful work into the issues surrounding AI integration, they conclude that "the task of knowledge acquisition has often been described as the chief bottleneck in the development of expert systems."[4] I know that the rest of this book will address the technical implications of this, but I would posit a different bottleneck: fear of change, and inversely the fearlessness that can lead to misuse. When reflecting on every stage of human history, we see this same clash of mindsets.

Often this embodies a conflict between traditionalists and reformers, between regulators and trailblazers. Instead of being seen as inevitable, I believe it is important to stand as twenty-first-century thinkers and develop this conflict thesis into one where the pros of each side can be used to benefit one another. This abstract may be too hypothetical in practice, but there is a lot of power in the idea of improving socio-political or academic hostility, even slightly. Indeed, if we take a further angle on this, I believe it to be the role of regulators to try to position themselves ahead of the game, so that development is regulated at a suitable rate before it becomes misused or dangerous.

Once the public notices the dangers of technology, as shown by the vast data privacy scandals occurring at the time of writing this book, then the sceptics begin to take arms and inhibit the otherwise productive use of disruptive technologies. This is also valid in reverse–that over regulation can stifle innovation. Ultimately, I believe a more holistic understanding would help smooth over competing viewpoints, in a way that identifies the human-centric positive values. In this context, a project called "Humanity 2.0," supported by the Vatican Dicastery for Promoting Integral Human Development, brings together financiers, philanthropists, artists, technology investors, politicians, and religious leaders to discuss a view towards the new destination that humanity is evolving to.

A DESIRE TO EXPLAIN THE UNEXPLAINABLE

Another point that I wish to address is the notion of intelligence, and the role of intelligence throughout the history of society. From a theological standpoint, we see that intelligence is often seen as split into two aspects (1) the physical knowledge gained and (2) the superior knowledge of a divine or metaphysical abstract. The former, based on sensory collection of information, creates the foundation for variations of a set of natural laws. The latter outlines explanations for events that lie beyond the feasible explanation.

This dynamic is also represented by Kant, with his description of the phenomenal and noumenal[5] realms that refer respectively to the physical and metaphysical aspects of life. The important point, however, is that throughout history, humanity has been focused on engaging with this exterior noumenal. Even technology has not managed to reduce the human desire to explain the unexplainable. This point is relevant today, at a stage in history where a huge innovative breakthrough attempts to explain the natural world. I'd like to look back at the Industrial Revolution again, standing as one of the only comparatively disruptive series of events to affect Western Europe from 1760 to 1840. Although many stopped believing in the noumenal and turned their belief to mechanisation, it also gave birth to a new set of thinkers that looked beyond the natural world. This is ultimately cyclical–the technologies that are incubated in universities get matched with world views that are reactionary in those very same spaces. For example, the Oxford Movement in the nineteenth century consisted of a crop of talented theologians and philosophers who all attended to and grew their theories while in Oxford. This movement, spearheaded by John Henry Newman, who will be canonised in October 2019, shows an awareness of life beyond the natural world and how even the most aggressive forms of innovation will keep people inquisitive.

This leads me to believe that humans will always have a desire for intelligence beyond the explained. This intelligence is crucial in balancing society with the outbreak of society, a desire that shouldn't be forgotten in an attempt to "not be outdated" but rather one that should be respected as a crucial aspect of humanity. Indeed, I would posit that this will be the balancing factor that helps put all innovation into perspective and allows humans to remain grounded. This is not to say that religion is the only way to activate this aspect of the brain, but rather it is just an example of one of the many ways in which humans can separate from the intense strains of the phenomenal world.

Therefore, I would argue that the theory put forward by Howard Gardner and Thomas Armstrong, stating that there are eight types of intelligence[6], is valuable for the implementation and construction of AI systems to operate the natural world. However, I would argue that there is another form of intelligence, one for the seemingly unintelligible and potentially theological, which history shows is at the centre of the human psyche and needs to be respected going forward.

MACHINE OVER HUMAN?

Changing tack slightly, while maintaining the tone from the previous paragraphs, I'd reverse the question and examine whether theologians and philosophers should be worried about AI, rather than the ways in which these disciplines can help the implementation of AI. The ever-increasing presence of complex artificial intelligence poses threats to all aspects of life, with a possibility of technological replacements for many facets of society. Notably influential figures in this field include the household names Bill Gates and Elon Musk, who state that the improvements in technology will, undoubtedly, cast a large shadow on humanity, a process that could take place within the next quarter century. An example of this is the agreement between Google and the New York City Council to implement driverless taxis, a huge leap in technology that leaves a sense of foreboding for the future dominance of machine over man. However, can the presence of artificial intelligence shake the age-old tradition related to theology, in the same way artificial intelligence is changing the taxi industry, to name one industry of many? The largest issue is whether artificial intelligence can truly become identical to humans–a debate that is essential to understand in order to judge the future of theology.

There are many aspects of humans that separate them from an inanimate object, or even another animal. Stemming back to Aristotelian ways of thinking, personhood is defined by the presence of a rational will. This helps place a human in both social and temporal contexts. This allows one to take into account the situation through the former context and any history or experience through the latter. The rational will is only present in humans, differentiating from carnal and vegetative wills present in other animals and plants, respectively.

Thomas Aquinas (1225-1274) adapts this Greek outlook by adding a religious angle to the presence of rational will. He argues that the rational will is only present because humans are created *imago Dei,* as said in Genesis 1:25. Thomists and Christians alike believe in the Sanctity of Life theory, which develops from Aquinas' teaching, stating that life is holy and sacred, with each human deserving respect for the distinctive and unique quality that is given by God.

An interesting adaptation of artificial intelligence that could affect theologians is that of technology providing a superior source of religious advice. For example, private confession in a church originated in 1564 when Cardinal Borromeo implemented the privacy of a grill between the priest and penitent. This tradition could, potentially, be adapted to include technology, which automatically gives the best advice to the penitent. Through deep-learning developments, such as Generative Adversarial Networks, this could even be taken further so that the face of your local clergy could be giving you automated and generated advice. Certainly frowned upon by the most conservative of churches, however, technology to achieve this capability of ethical discussion is in the early stages of development and could be seen in more liberal denominations in the future. The real question lies in the system's ability to convince the penitent that it could be a human priest giving the advice, or rather that the advice given is more knowledgeable than that of a human priest's. The Turing Test examines a computer's intelligence through communication, originally devised by one of the leading lights in twentieth-century computer science and World War II code breaker, Alan Turing. Recent discoveries through this test reveal that developments in computer technology have allowed for humans to be "fooled" when engaged in a basic text conversation. For example, there is no audible speech simply text generated by the computer in answer to the text typed by the human being tested. An example is at the Royal Society in London where an event was held to test the computer programme called Eugene Goostman, an artificial intelligence software pretending to be a 13-year-old boy. It tricked 33 percent of the expert judges into thinking it was a real boy during a five-minute text conversation. This shows the rapid improvement in communications technology.

If the basic technology exists to produce believable answers, then the next step is to create a smooth operating voice to make the text audible. The logical next step, using technology such as GANs (more on GANs in Chapter 2), would be to turn this text into both vocal and visual manifestations.

CAN MATH BE USED TO CODE ETHICAL BEHAVIOUR IN TECHNOLOGY?

Certain forms of ethics can be considered mathematical and systematic, a procedure that includes using formulae or equations to judge the correct moral action. As Spinoza argues, "I should attempt to treat human vice and folly geometrically . . . the passions of hatred, anger, envy, and so on, considered in themselves, follow from the necessity and efficacy of nature. . . . I shall, therefore, treat the nature and strength of the emotion in exactly the same manner, as though I were concerned with lines, planes, and solids."

The quote above by, arguably, one of the most influential philosophers, taken from his *Magnum Opus,* states the mathematical importance in even the Christian forms of ethics. Furthermore, utilitarianism can be seen as similarly systematic through its three main branches presented by Jeremy Bentham, John Stuart Mill, and Peter Singer. Fundamentally, Bentham's form of utilitarianism states a very quantitate form of analysing ethics by use of a Hedonic Calculus, a rating system that declares which actions should be deemed as moral. Mill's utilitarianism states the importance of higher or lower pleasure. This qualitative approach is able to distinguish between "good and bad" pleasures, and from this he states that "it is better to be a human being dissatisfied than a pig satisfied."

Lastly, Singer's utilitarianism states the importance of sentience when deciding pain and pleasure, categorising different items into "tiers" depending on sentience. These three forms of utilitarianism allow a clear pathway for technological aid in decision making, using mathematical formulae or code to create the indisputably correct moral action. This could impact the future of theology as decision making becomes less and less debatable, furthermore questioning the God-given rational sense argued by Christian ethicists that is the sole reason we are able to act correctly. This could result in the future of people having more faith in technology than in God.

Immanuel Kant's ethics are based on three formulae: Universality, Humanity, and the Kingdom of Ends. These help construct a categorical imperative, free from *Neigungen* (inclinations), from which maxims (moral laws) are formed. In order to test the morality of these maxims, Kant proposes two tests, in which the true applicability of the maxims is examined. This system is very systematic and could also easily be applied to a computer's software, making technology of a superior moral standing than humans. This point, perhaps, produces an image of a horrific dystopia, however, it is important to take into account the ability of artificial intelligence to become ethical beings.

Clara Durodié

WILL AI ELIMINATE RELIGIOUS BELIEFS ?

A subsidiary point of the one above is that, on a basic level, artificial intelligence will help the development of science, a discipline that continues to disprove certain aspects of religious faith. Artificial intelligence, therefore, indirectly influences the future of theology, as science will leave less and less to faith and more to rationally coherent concepts. The future, one could argue, could develop into a society that agrees with the eighteenth-century philosopher Gotthold Ephraim Lessing's proposition about the knowledge of Jesus and God. The principle was later named *Lessing's Ditch*, stating that he could not make the jump from rational science to belief in God's metaphysical "magic." He proceeds to say that religious claims are merely historical claims (especially such leaps of the imagination as Jesus' resurrection). He names this distinction as "the ugly broad ditch which I cannot get across, however often and however earnestly I have tried to make the leap." One could argue that this could be a representation of the future approach towards religion: perhaps Jesus existed as a historical figure, however, nothing more than that due to the scientific power to disprove any other claims. One who contradicts certain scientific proof against religious miracles could fall into the trap of fideism (the blind faith that ignores reason). Ultimately, it is undeniable that artificial intelligence will affect the art of theology and philosophy, as it will influence all aspects of everyone's life, whether directly or indirectly. However, by no means do I suggest that religion will be made extinct, as faith is able to be present even in the face of ever-expanding scientific knowledge. Furthermore, artificial intelligence will never truly take the form of an unrecognisable human due to the intrinsic and inimitable nature of human creativity. The idea of creativity is not *just* a view; it is a psychologically ingrained nexus of perception and motivation that indisputably makes humans distinctive enough not to be able to be artificially reproduced. Therefore, the intuitive nature of humans will remain unchanged. But will theology will remain fundamentally unchanged? Or not?

MACHINE INTELLIGENCE AND HUMAN INTELLIGENCE

Humans' fascination with intelligent artefacts dates back to 1637 when Rene Descartes, the French mathematician and philosopher, stated that it will never be possible to build a machine that thinks the way humans do. According to him, machines think through their electric circuits and components not through their thought. About 200 years later, in 1840, Ada Lovelace was the first to recognise the full potential of a computing machine, and she was the first to write an algorithm to be used on such a machine. She is one of the first computer programmers. Inspired by his work with Ada, Charles Babbage proposed a mechanical general purpose computing machine. On this work, 110 years later, in 1945, Alan Turing proposed the question "Can machines think?" in his now famous paper. And finally, about 70 years later, we are talking about giving robots legal personhood, and trans humanism proposes that we expand humans to reach a trans-human state, allegedly superior to that of human intelligence.

BRAIN AND MIND

AI impacts people. As the industry is being told to stay ahead of the game to "disrupt" and "innovate," we need to be mindful and predict the impact this technology might have. And there is no path to it, unless basic theological and philosophical concepts are understood and applied. This is why the rest of the chapter tries to summarise core concepts that might help you understand why financial services practitioners and board directors need to be vaguely familiar with the humanist understandings of intelligence. The nexus of many discussions on intelligence is driven by two different schools of thought:

1. Dualism asserts that there is a distinction between mind and body. Our mind is more that just our brain, it has a spiritual dimension, which covers consciousness, spirit, and soul, and it is our non-tangible self surrounding our tangible self (our body). The non-tangible self manifests through our tangible self.
2. Materialism asserts that everything is made from physical materials and has physical properties. It contends that we are entirely made of matter and energy, just like the rest of the universe. This is why we are irreversibility connected to that universe. Constructal law,[7] a new evolutionary theory across the board, explains that for anything to evolve it has to have unrestricted access to what makes it evolve, and that includes intelligence.

Humanism vs. Transhumanism

These are two schools of thought with fundamentally different agendas; both have rich and complex histories and narratives. They have sparked incendiary discussion and disagreements. Humanism is inspired by technological developments and is seeking to apply them to create a world that fits human needs and supports humans to live a better life. They put human-positive values at the center of the development of any form of technology. They are the likely promoters of responsible AI, demanding explainability and auditability of systems. On the other hand, Transhumanism promotes an agenda that leaves humans behind by using technology to create new forms of existence that transcend the limitations of the human condition. They rely on NBIC technologies (nanotech, biotech, information technology, and cognitive sciences) to achieve this transcendence. They view these technologies to be more or less integrated in human bodies. Transhumanists are the likely promoters of legal personhood for AI, thus creating indirectly a thick legal veil that would protect bad players.

Intelligence

We continue our journey in the space of no clear agreement on definitions. Intelligence is yet another concept which has many definitions, because intelligence means many things to many people. This means that there is not a definitive definition of what intelligence is and also there are competing definitions like the capacity for introspection, presence of consciousness, self-awareness, creativity, capacity for logic, understanding, planning, problem solving, and learning.[8] However, in 1982 two American professors, Robert J. Stenberg and William Salter, arrived at what I believe is the most suitable definition: "Intelligence is goal-directed *adaptive* behaviour."[9]

There a few more definitions that come close:

- ability to accomplish complex goals (Tegmark, 2017)
- ability to acquire and apply knowledge and skills (Oxford Dictionary, 2018)
- ability to learn, understand, and make judgments or have opinions that are based on reason (Cambridge Dictionary, 2017)
- ability to solve problems and attain goals in a wide variety of environments (Murray Shanahan, 2018)

The transition from intelligence to artificial intelligence is grounded on the core assumption adopted by computer scientists – the human brain is the blueprint, the training ground from which they can learn how to build systems that mimic the brain. But this is not that simple because there are still many unknowns about how the brain works. The recent advancements in deep learning, which have propelled business applications in narrow AI in financial services, are based on designing artificial neural networks that mimic the brain. Some work very well and have achieved human parity in a range of tasks like voice and object recognition. But most of them lack context. Intelligence is not a single dimension, and general purpose intelligence, artificial or human, does not really exist in humans, and it will be heavily cost constrained in artificial mediums trying to emulate the human brain.

Agency and Free Will

Breakthrough developments in AI will continue to raise piercing questions about free will and agency. I recommend a seminal discussion between Dr Fei-Fei Li and Yuval Harari at the opening of The Stanford AI Center. Yuval Harari made four comments that are valuable points to reflect on:

- "Who decides what is a good enhancement and what is a bad enhancement? So how do you decide what to enhance if, and this is a very deep ethical and philosophical question–again that philosophers have been debating for thousands of years–[we do not have an answer to the question] "what is good?" What are the good qualities we need to enhance?
- If you can't trust the customer, if you can't trust the voter, if you can't trust your feelings, who do you trust?
- What does it mean to live in a world in which you learn about something so important about yourself from an algorithm?
- The engineers won't wait. And even if the engineers are willing to wait, the investors behind the engineers won't wait. So, it means that we do not have a lot of time," Harari warned that we need policies in place for this fundamental change he refers to as Hacking Humanity at scale.

Understanding the meaning of agency and synthetic agency in technology provides a new lens to better understand the ethical implications of using this technology in designing and providing financial services. This enables us to understand the dramatic importance of ensuring human involvement and supervision of AI systems, the organisational risk of anthropomorphising AI systems (referring to them as "she" or "he" rather than "it" or attributing them human-like abilities or visual descriptors) as professor Joanna Bryson demonstrated in one of her papers that presenting robots as people stops us from thinking clearly about AI.[10] The principles of agency and autonomy in artificial intelligences need to be well understood and so do the risks associated with using this technology.[11]

The concept of freedom of choice or free will is a philosophical conundrum retail banking will be faced with–shall we let customers decide what's best for them or shall we use AI for gentle nudging and benevolent manipulation of choice like in the case of gambling addicts or self-destructive spending patterns? Moral laws are good guiding principles to use.

I would recommend Margaret Boden's books, as her work provides useful and detailed insights, especially *Mind as Machine: A History of Cognitive Science.* Nils J. Nilsson's easy to read *Understanding Beliefs* is a concise introduction to philosophical angles of how people come to know things, therefore we understand how machines know things about their surroundings. For my readers who would like to have an advanced introduction to wider connected concepts, I would recommend *The Cambridge Handbook of Artificial Intelligence* edited by K. Frankish and W.M. Ramsey.

Emotions and Machines

Humans are driven by emotions. We experience emotions, which in turn drive a range of reactions and actions. Our emotions exist because our minds learn from the environment, but more important, they are a reflection of our own self. "I do think there's value for people in understanding how emotions and emotional states play into our perception of the world and underpin our judgements and decisions. Emotions themselves are also perceptions–but of the self rather than of the world"[12] is a valuable explanation from Anil Seth, British professor of Cognitive and Computational Neuroscience at the University of Sussex.

Our minds search for knowledge in an attempt to optimise or explore the information we have available. In the same way, algorithms are designed to achieve the same. And just like our minds, which mix and match information, algorithms can be mixed and matched to optimise how the desired outcome is achieved. For example, the Google Deep Mind's Alpha Go combined a rich mix of algorithms of different techniques (Deep Learning, Monte Carlo Tree Search, and Reinforcement Learning) to achieve a historic win against the best player of Alpha Go in 2016, a win that went on to shape the AI Moment (more on this later). Machine Intelligence has no common sense, hence it is unable to detect emotions and make sense of basic situations.

On Consciousness

The Oxford Dictionary defines consciousness as "The state of being aware of and responsive to one's surroundings."

Sir Roger Penrose, a distinguished mathematics physics professor at the University of Oxford, who, amongst many achievements, won the illustrious Wolf Prize for physics in tandem with Stephen Hawking, wrote in 1995 that "Consciousness is part of our Universe, so any physical theory that makes no proper place for it, falls fundamentally short of providing a genuine description of the world.

There is no physical, biological or computational theory that comes very close to explaining our consciousness and consequent intelligence[13]." "I cannot imagine a consistent theory of everything that ignores consciousness."[14] And for a topic that permeates everything, it is the least understood, the most controversial, and the most philosophically difficult concept.

Consciousness is "controversial" because it is one of the "thorniest philosophical topics of all," and it is perhaps the least understood and it is linked to the field of artificial intelligence because of the need to "predict which intelligent entities have subjective experiences."[15] Some experts believe that in the absence of a thorough understanding of where and how consciousness exists, we should refrain from using it as the deciding factor to confer legal rights on machines.

Yet, others are keen to confer legal rights on machines precisely because we do not know what consciousness is, so they feel we need to make sure that we do not deplete those machines from their rights. The latter category blindly opens a dangerous possibility for bad players to evade legal responsibility for the machines they built. This book warns against such practices.

The discussion around consciousness in connection with AI has three main drivers:

1. Feelings: How does it feel to be an algorithm homed either in a smart device (iPhone) or in an autonomous one (Tesla autonomous vehicle or Boston Dynamics robot)?
2. Legal rights: Is consciousness the litmus test to enshrine legal rights for AI systems (static or autonomous machines)?
3. Freedom: Is it for those who want to upload their minds to break free from biological limitation as transhumanists suggest?

Daniel Dennett is one of the contemporary philosophers whose work I warmly recommend, primarily because I feel it resonates with my thinking. I would recommend two of his books *Consciousness Explained* and *From Bacteria to Bach: The Evolution of Minds.* Murray Shanahan's *The Technological Singularity* is another recommended book. I also draw guidance from the work of the philosopher John R. Searle. I recommend the following books by him: *Intentionality: An Essay in the Philosophy of Mind, Seeing Things as They Are: A Theory of Perception,* and *Speech Acts: An Essay in the Philosophy of Language.*

Moralising Technology

Technology already exists in nearly every layer of our society, and as it permeates deeply into financial services, it shapes our lives and how we think and engage with the world around us. We are yet to understand the long-term neurological and physical development implications.

What we have yet to grasp is the moral impact of the technology that we embed in financial products, invest in, and put forward to our customers.

"Maybe even more difficult for computers, but also quite important, will be to understand not just human emotions, but also something a little bit more abstract, which is our sense of what's right and what's wrong," warns Yoshua Bengio, a Canadian computer scientist, most noted for his work on artificial neural networks and deep learning. So, we cannot delegate this function to computers–that would be a fallacy. In his book *Moralizing Technology,* Professor Peter-Paul Verbeek offers a framework of reference to evaluate the value of new inventions and of the digital financial products that we promote to our clients.

KEY POINTS TO REMEMBER

- It is crucial to understand the impact that groundbreaking inventions have had on the past, to better understand the future.
- Once the public notices the dangers of technology, then the sceptics begin to take arms and inhibit the otherwise productive use of disruptive technologies. This is also valid in reverse–that over regulation can stifle innovation.
- Humans will always have a desire for intelligence beyond the explained.
- The largest issue is whether artificial intelligence can truly become identical to humans–a debate that is essential to understand in order to judge the future of theology.

CHAPTER TWO

The Taxonomy of Artificial Intelligence

This chapter covers the fundamental concepts and definitions of AI technologies. These are basic concepts needed to understand what AI is, why these technologies are different, and when to use them. I'll take you through the different AI approaches, or "tribes," an overview of AI technologies, common misconceptions about AI, and some of the key terminology. AI consists of generated *predictions* based on mathematical models. Different algorithms are used to solve different problems, and sometimes they need to be mixed to find a richer solution to address a problem.

Machine learning (ML), a subset of artificial intelligence, is a term that you will see repeatedly throughout the book, and this chapter will thoroughly explain how it works and why it is important for delivering AI. It will also outline the three types of machine learning–supervised, unsupervised, and reinforcement learning–and their challenges and applications. It will cover *deep learning* (DL) as well, which is a subset of machine learning that attempts to simulate human neural networks in computer programmes, creating artificial neural networks. Within deep learning, we'll cover convolutional neural networks, capsule neural networks, recurrent neural networks, and generative adversarial neural networks. We'll examine other approaches to machine learning such as semi-supervised learning, transfer learning, federated learning, genetic algorithms, and even machine "teaching." Finally, we'll explore natural language processing (NLP), computer vision, expert systems/rules engines, and types of computing. Once you have a basic understanding of how these approaches work, what differentiates them from one another, some of their core challenges, and how they are being used or applied, you will have a better sense of the types of AI that may be useful for your business. I also recommend further reading to understand the inner workings of the frequently used methods like backpropagation, stochastic gradient decent and their mathematical underpinnings. As tempting as it was, I made an effort to stay away from mathematical equations or indeed develop on linear algebra concepts. For this, a good starting point might be the latest edition of Artificial Intelligence: A modern Approach by Stuart Russell and Peter Norvig.

DEFINITIONS OF AI, AI TRIBES, AND AI APPROACHES

In essence and to oversimplify complexity, *AI is a software (computer program or algorithm) that generates predictions based on mathematical models.*

AI is a multidisciplinary field that gathers input from mathematics, statistics, physics, computer science, philosophy, sociology, biology, neuroscience, and psychology. Each discipline has a different perspective on how specialists believe that learning in machines should be built. These distinct approaches have become the differentiator of "AI tribes." The Tribes are essentially defined by what experts believe is the best way to solve a problem, or a goal. Y Combinator (a world-leading US technology incubator) talked about 21 different AI cultures/tribes at the 2006 IEEE Conference (Institute of Electrical and Electronics Engineers, the world's largest technical association) and identified the top 10 machine-learning approaches (tribes).

One of the leading researchers, Pedro Domingos, identified the five approaches or methods to machine learning in his book *The Master Algorithm*. He calls these the Tribes. Domingos posits that these methods can be combined to make the algorithms that will redefine our world. Some would refer to it as AGI (Artificial General Intelligence).

Table 2.1 AI Tribes

	TRIBE	ORIGIN	MASTER ALGORITHM
1	Symbolists	Logic, philosophy Focus on philosophy and psychology, as ways to experiment in order to learn	Inverse deduction
2	Connectionists	Neuroscience Attempt to mimic human brain	Backpropagation
3	Evolutionaries	Evolutionary biology is the foundation of learning in machines	Genetic programming
4	Bayesians	Statistics Focus on statistics and probabilistic inference	Probabilistic Inference
5	Analogizers	Psychology Focus on the intersection of mathematical optimisation and psychology to extrapolate similarity judgements (Ahmed, 2017)	Kernel Machines

There are also different definitions of AI that correspond to two main approaches:

- *Human-centred Approach:* based on observation and hypothesis about human behaviour
- *Rationalist Approach:* based on a combination of mathematics and engineering

The various groups "have both helped each other and disparaged each other"[16] because they have never agreed on core definitions of intelligence, or what it means to be human. These approaches are summarised below:[17]

Table 2.2 Definitions of AI

THINKING HUMANLY	THINKING RATIONALLY
The automation of activities that we associate with human thinking, activities such as decision making, problem solving, and learning. (Haugeland, 1985)	The study of the computations that make it possible to perceive, reason, and act. (Winston, 1992)
ACTING HUMANLY	ACTING RATIONALLY
The study of how to make computers do things at which, at the moment, people are better. (Rich and Knight, 1991)	AI is concerned with intelligent behaviour in artifacts. (Nilsson, 1998)

Narrow, Broad, General, and Super Artificial Intelligence

Other definitions of AI use human intelligence to benchmark algorithms' capacity to prove a level of intelligence that is either narrow, broad, general, or super:

- Narrow Artificial Intelligence (Narrow AI): This refers to a specific process, field, or function (e.g. human resources, legal function, marketing, or portfolio management). This type of AI is all there is currently in production and use.
- Broad Artificial Intelligence (Broad AI): This is the next-evolved version of Narrow AI, and unlike Narrow AI it is able to learn and deal with the next adjacent domain to its core learning.
- Artificial General Intelligence (AGI): This type of AI is not confined to one specific process or domain, but it acts in a human-like cognitive way to achieve goals across multiple domains.
- Artificial Super Intelligence (ASI): Super-human AI is a concept introduced and promoted to the public by one academic philosopher, David Chalmers.[18] Talking about ASI with so much conviction has captured everyone's imagination, for better or

worse. However, closer to the practical reality of the possibility of building ASI, computer scientists and researchers believe that human intelligence cannot be surpassed. And yet, some insist[19] that ASI will take the form of general reasoning systems, which will operate in a dissimilar way to human cognition, therefore there will be no limitations on what ASI can do. It will, according to speculation, also act in its own best interest, which will make it unstoppable in achieving this goal. This thinking works on the assumption that ASI's goal would not be aligned with humanity's goal. In a moment of clear thinking, scientists remind us that it is possible to achieve this alignment as long as three main problems are solved: "making machines (1) learn our goals, (2) adopt them, and (3) retain them."[20]

Artificial Intelligence Includes Machine Learning

"Artificial intelligence (AI)" and "Machine Learning (ML)" are two terms that are *incorrectly* used interchangeably. ML is a subset of AI, *but* AI ≠ ML, because AI is more than ML. Some people claim to be experts in "Artificial Intelligence *and* Machine Learning." To some this is a mouthful and sounds impressive, but when you unpack it, it's like saying in *one* sentence "I like dogs *and* golden retrievers." Because AI and ML have been used incorrectly, the confusion runs deep and has spread like wildfire throughout conferences and self-proclaimed AI experts in financial services, but my reader will be well advised to see these errors.

In addition to the above, here are examples of such chaotic statements:

- "Machine learning makes artificial intelligence possible."[21] No, it does not. Machine learning is a subset of the artificial intelligence field. It's like saying "the steering wheel makes a car possible." The steering wheel *is* part of a car, and yet it does not make the car possible.
- "Phases of artificial intelligence: machine learning, deep learning, and reinforcement learning." These types of learning are not a phase of artificial intelligence; they are AI tools or techniques. You will see below a clear description of how these terms interrelate.
- "AI augmentation will fuel net job growth by 2020."[22] There's no such a thing as *AI augmentation.*
- "AI and machine learning will transform wealth management." A marketing expert and contributor to *Forbes* proclaimed in her piece that debuts with fear mongering: "Every minute of every hour we hear more about the vast, sometimes scary power of

artificial intelligence." Then she changes her tune and moves into merely listing a soup of terms: "And technology can do incredible good. When technologies such as AI, machine learning and natural language processing (NLP) are used not to replace people [....]" [23]

Some Basic Terminology Used in AI

Before we proceed to explain the most important AI techniques, I believe that clarification of misappropriate terminology is necessary. In the interest of *intellectual hygiene,* we need to shine the spotlight on the most-used terminology, explain it, and so help everyone to make themselves better understood.

- AI and Machine Learning shouldn't be used in the same sentence. It is factually wrong. Machine leaning is a subset of AI, so it is pointless to use them both in the same sentence.
- Robots are the physical embodiment of AI technologies–they can be built to see, to talk, to hear, and to move. This may imply physical and decision-making autonomy. It is incorrect to use "bots" when referring to a software or digital approach like "Robots are doing the investments" or indeed "robo-advisers." Use digital instead: *digital adviser*.
- Computers or Machines are your desktop, laptop, or any smart device that hosts an algorithm either permanently or temporarily. Sometimes algorithms and machines are used interchangeably.
- An Algorithm is a piece of software that is programmed using techniques under the AI umbrella with the objective to:
    - learn either on its own or assisted by a programmer
    - provide predictions, insights, and classification
    - make autonomous decisions
- AI, AI systems, or AI tools mean different types of algorithms.
- Bot or bots is a software that is built using Robotic Process Automation (RPA) techniques and is referred to as linear automation.
- Linear automation is a software that is programmed to achieve a basic task and is given a set of specified rules to abide by. The software is not able to provide any insight on its work and does not learn from its work. Sometimes, it is referred to as "dumb automation." Some of the linear automation tools can be upgraded to exponential automation.

- Exponential automation is a software that is programmed using AI technologies; it provides insights and learns from its work.

AI Research Focus

AI research is focused on five fields of analysis, which are impeccably synthesised by David Kelnar in *The 2019 State of AI* report and are reproduced below with permission:[24]

1. *Knowledge*: The ability to represent knowledge about the world. For software to possess knowledge, it must understand that certain entities, facts, and situations exist in the world; these entities have properties (including relationships to one another); and these entities and properties can be categorised. Tasks: information synthesis, recommendations, consumer segmentation
2. *Reasoning*: The ability to solve problems through logical reasoning. To reason is to apply logic to derive beliefs, related ideas and conclusions from information. Reasoning may be deductive (derive specific conclusions from general premises believed to be true), inductive (infer general conclusions from specific premises) or abductive (seek the simplest and most likely explanation for an observation). Tasks: legal analysis, portfolio management, application processing , compliance, risk management
3. *Planning*: The ability to set and achieve goals. For software to be able to plan, it must be capable of specifying a future, desirable state of the world and a sequence of actions enabling progress towards it. Tasks: logistics, predictive maintenance in trading algorithms, demand forecasting,, actions optimisation
4. *Communication*: The ability to understand written and spoken language. To communicate with people, software must have the ability to identify, understand and synthesise written or spoken human language. Tasks: voice control, conversational AI, customer support, real-time transcription and translation, investment management.
5. *Perception*: The ability to make deductions about the world based on sensory input. To perceive, software must be able to organise, identify and interpret visual images, sounds and other sensory inputs. Tasks: autonomous vehicles, authentication, augmented reality, surveillance, industrial analysis - drones.

How to Evaluate Algorithms

There are best practice software engineering guidelines which, among many other details, show that implementation should be well designed and documented.[25] The starting point in the analysis and evaluation of an algorithm is the implementation model, which includes details on the resources of that technology, efficiency of the model, and costs. Think of building a house. You need to list (model) the resources and costs you have to build a house design. You'd then be able to evaluate and analyse if that house design is suitable for what you want to achieve. Efficiency to achieve the task at hand, like anything else in life, is a good starting point to evaluate how the algorithms work and what to choose. The rule on algorithms is that "different algorithms devised to solve the same problem" would do so with different levels of efficiency.[26]

ARTIFICIAL INTELLIGENCE TECHNOLOGIES: AN OVERVIEW

AI technologies form a vast and complex field. The diagram below helps visually understand the most important AI subsets, how they are correlated, and what tasks they perform.

Figure 2.1 AI Technologies

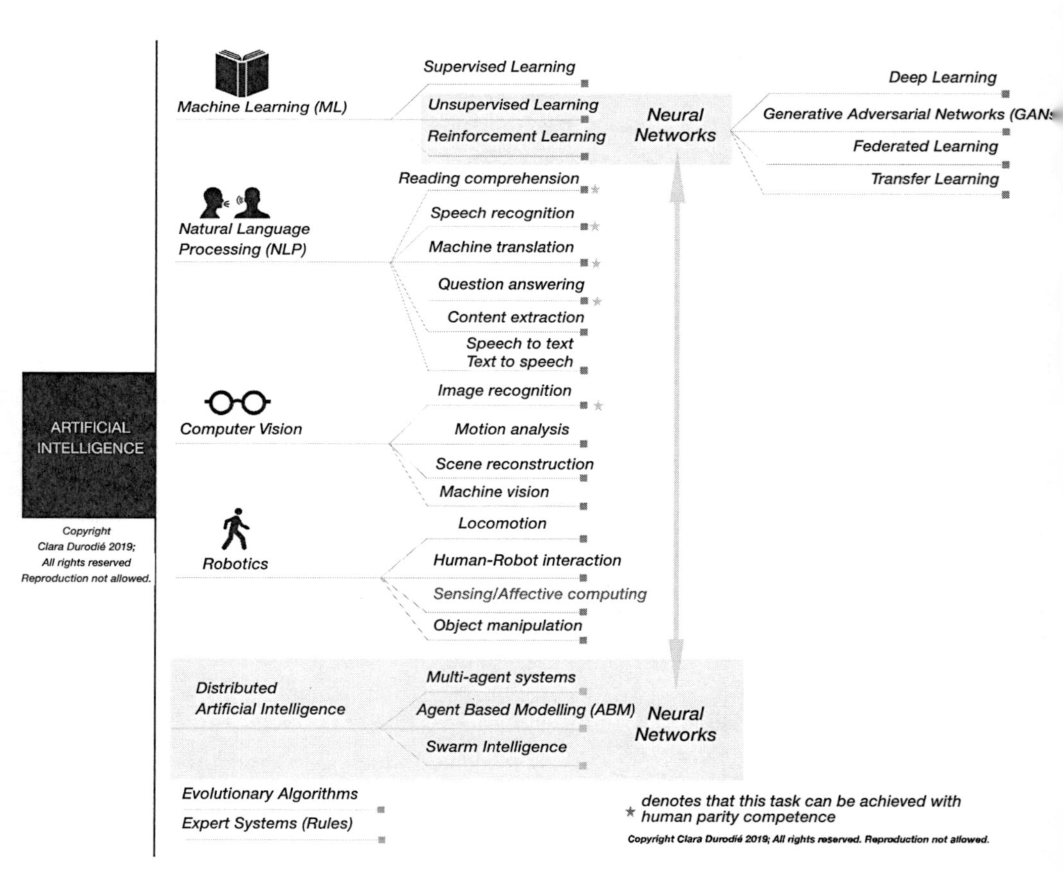

Machine Learning (ML)

Definition: Machine learning is a subset of artificial intelligence, which is part of the computer science field. Like with humans, the ability to learn is the foundation of intelligence. And so, researchers have spent considerable time identifying the best ways for machines to learn, hence the term machine learning. Learning is done in advance and during the delivery of a solution to any problem. Learning presumably comes from experience, practice, or training, not solely from reasoning. Machines learn by extracting patterns from data, which can be any form: video, text messages, likes on social media, temperature reading, interest rates, utilities expenses, credit card transactions, or stock trades–just about anything can be captured, measured, or represented in digital form.[27]

Machine learning is a term coined in 1959 by Arthur Samuel. It uses statistical tools or mathematical models to enable machines to improve their performance without dedicated programming. So, *ML algorithms* learn on their own what's the most optimal course of action to achieve a task or solve a problem.

Challenges[28]:

- Overfitting: One of the biggest challenges in ML, it means that the model/algorithm becomes very good with the learning on the training data to the point that it does not understand what its knowledge means in the context of new data.
- Underfitting: This means that the model is neither very good with learning from the training dataset nor with extrapolating what it learned to new data.

Figure 2.2 How Do Machine Learning Algorithms work?

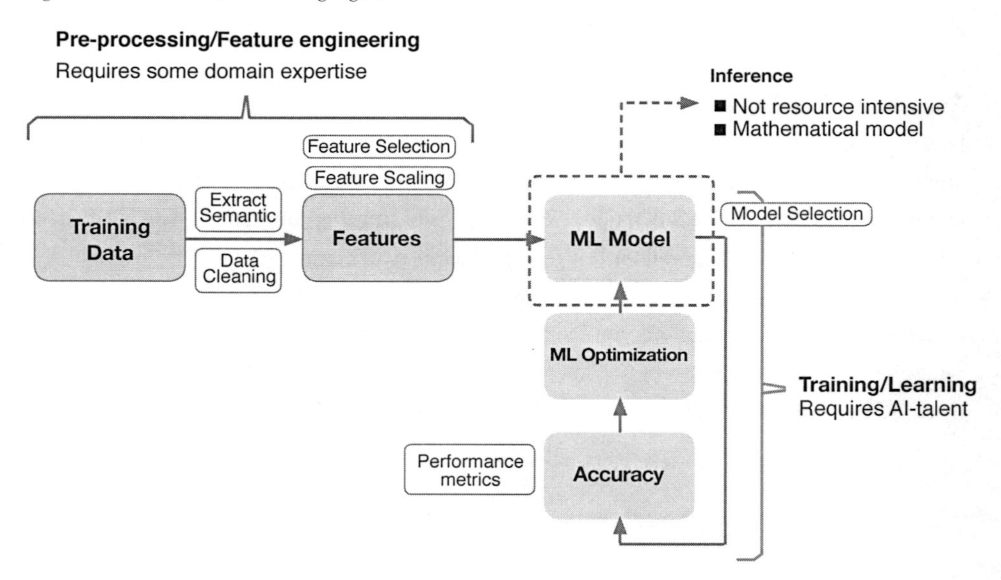

There are a few different machine-learning approaches and I recommend the following categorisation–supervised learning, unsupervised learning, reinforcement learning, and deep learning. Each has its own challenges and applications, so it's important to understand their differences when you are planning to incorporate ML into your business strategy.

Machine Learning Approach 1: Supervised Learning (SL)

Definition: SL is a subset of machine learning that works with labeled data and makes predictions based on a given set of examples, so there is a predetermined outcome. The goal of supervised learning is to optimise the mapping of a set of inputs to their correct output. For example, the classification goal has the input (given set of "variables") as a set of animal images, and the correct output is the type of animal (dog, cat, zebra, etc.). In other words, each type of animal has been assigned a label to help the model match it with what it finds in the "variables" pool. Algorithms spot trends in historical data and use that to make predictions when they are presented with new data.

When the goal is to predict a specified variable, then it is also referred to as *classification*. There's a range of algorithms that are used for classification: Support Vector Machines; Decision Tree and its more complex relative, Random Forests; Naive Bayes; and Nearest Neighbour. When more than two labels are used, then there's a *multi-class classification*.

Challenges: Data needs to be labelled for the algorithms to understand what it is and how to deal with it. When data is not labelled, it is not seen and so it cannot be used. Supervised learning is expensive because (1) data labelling is time consuming and (2) requires large datasets.

Applications: Traditional approaches like linear regression and logistic regression are bedrock algorithms that have been used extensively by banks in risk management applications to make predictions in modelling insurance underwriting, credit default risk, and fraud detection. In fraud detection with supervised learning, the sample data is labelled "fraudulent" and "non-fraudulent." The big challenge in fraud detection is to correctly identify fraudulent transactions so that legitimate transactions are not declined. For this reason, in certain circumstances semi-supervised learning algorithms offer a more accurate outcome, as they enable access to more data (unlabelled) and therefore it fights fraud dynamically as cyber criminals or attacks change their patterns of actions.

Machine Learning Approach 2: Unsupervised Learning (UL)

Definition: This approach to machine learning works solely with unlabelled data and makes *predictions* on the patterns it finds, so there is no pre-determined outcome. This approach to learning is similar to how humans learn and solve problems. It is an important approach for understanding variation and grouping in unlabelled data. The most popular algorithms are K-means, Hierarchical, Gaussian mixture, Graph Theory, or a mix of different learning algorithms.

These algorithms are given unlabelled data. They are asked to discover patterns and describe hidden distribution or structure of data. The goal is *clustering* the data, meaning to group data in sub-groups that have a higher level of similar characteristics, e.g. investment risk appetite. Clustering or segmentation might be more relevant when done within the sub-groups rather than the whole dataset. The number of variables can be as vast as the sub-groups of data. Sometimes, the size of the dataset might bring too much *noise* in the data. Depending on the goal, it might be necessary to experiment with reducing the size (dimension) of datasets and by implication reduce the number of variables. This is useful when some variables are identified as not relevant to solving a problem. This approach is called *dimensionality reduction*, and it helps with achieving more clarity on specific correlations, interconnections, and true relationships. In simple terms, it's about seeing the forest for the trees. Computer scientists and researchers are focused on using unsupervised learning to bring *deep learning* developments to scale while achieving a high level of outcome precision.

Challenges: Unsupervised learning is more subjective than supervised learning, as there is no simple goal for the analysis, such as the prediction of an output.

Applications: Customer segmentation from a range of unstructured data like demographics, web crawling history, what they like or dislike, purchase history, or feedback. This is useful to build models that give customers personalised attention more than directing them to the nearest gym or hairstylist.

Machine Learning Approach 3: Reinforcement Learning (RL)

Definition: This approach to learning is based on (1) trial-and-error and (2) reward to achieve progress with its task. The algorithm is not given a training data set, so it learns from its own experience by exploring what happens when it does something. It forms its experience from feedback from the environments, such as "this action was good" or "this action was bad," and from human observations rewards as it attempts to achieve its task without hand-engineered intervention or domain knowledge (e.g. selling or buying a stock).

With RL, learning is experience driven rather than pattern based, which is the case in unsupervised learning. This means that the algorithm (agent) generates its own experience, otherwise referred to as its own training dataset. This translates in low cost for training data (in contrast with supervised learning, which is expensive to label the data). This is why RL is regarded as having a long-term impact on democratising access to artificial intelligence.

Reinforcement learning was refined by DeepMind in 2014 as Deep Reinforcement Learning, with the "first artificial agents to achieve human-level performance across many challenging domains" within the gaming environment (Atari and Go game).[29] It is called AlphaGo and it was the first computer program to defeat a world champion at the ancient Chinese game of GO. Alpha Go, which was then upgraded to AlphaGoZero, is not only "even more powerful and is arguably the strongest Go player in history"[30] but also leaned to play chess without a human-reward input. But why is AlphaGoZero groundbreaking? It works when human knowledge is not available, too expensive, or too unreliable; it is its own teacher; this type of *artificial intuition* has never existed in previous technologies.[31] It substantially improves performance because it uses one neural network that combines *policy networks* to select the next move to play and *value networks* to predict the winner from each position.[32]

Challenges: In computer vision, one difficulty modern reinforcement learning systems seem to have is knowing what parts of a problem to devote attention to exploring, so they do not waste time on less interesting parts of the image. This is usually referred to as *explore-exploit dilemma*.[33] Another challenge is the credit assignment problem, which means determining which action was responsible for *reward* and equally which one for *punishment*.[34]

Applications: There are developments that aim to demonstrate that this framework delivers results in various portfolio management strategies.[35] More work has been done to blend different frameworks beyond deep learning, like an assortment of "flavours." In one case, the authors claim a great achievement that "although with a high commission rate of 0.25% in the backtests, the model achieves at least 4-fold returns in 50 days" back-tested in a cryptocurrency market environment.[36] Another example is RL in individual retirement portfolio management, a use case that hasn't been explored before. The authors of this framework have used real-world datasets and have concluded that "a dynamic RL agent performs the best."[37]

Machine Learning Approach 4: Deep Learning (DL)

Definition: Deep learning, sometimes amalgamated with Deep Neural Networks or Artificial Neural Networks, is a subset of machine learning. DL is designed to learn inspired by organisational principles that neuroscientists identified in how the human brain works.However, the relationship between artificial neural networks and human neural networks remains aspirational.[38]

For many years, computer scientists thought they should take inspiration from how the brain works to try to make AI work better, by measuring how neurons spike while receiving a signal. However, they didn't really get very far–they made a push in the 80s, but then they stalled, and they were kind of laughed at by everyone in AI saying, "you don't look at a bumblebee to design a 747." But it turned out the inspiration they received from looking at the brain was extremely relevant, and without that, they probably wouldn't have gone in that direction.[39]

Computational neuroscience researches the actual structure of the human neurons and simulates these in a computer programme. Their ultimate goal is to learn how the brain really works. And yet, the biological brain remains "a riddle wrapped in a mystery, inside an enigma." It is no wonder, considering that a human brain has more than 80 billion neurons and tens of trillions of synapses to send and receive electrical or chemical signals. When the signals become unusual, a neuron "fires" or "spikes," meaning it lets the neurons it is connected to know about the unusual signal. One of the questions that remains unanswered is "why do neurons really spike?" This is a cardinal question for further developments in DL, and it could lead to smart AI systems that can store more information more efficiently, according to the English-Canadian Professor Geoff Hinton, often referred to as the "godfather" of deep learning.

Figure 3.3 How Do Neural Networks Work?

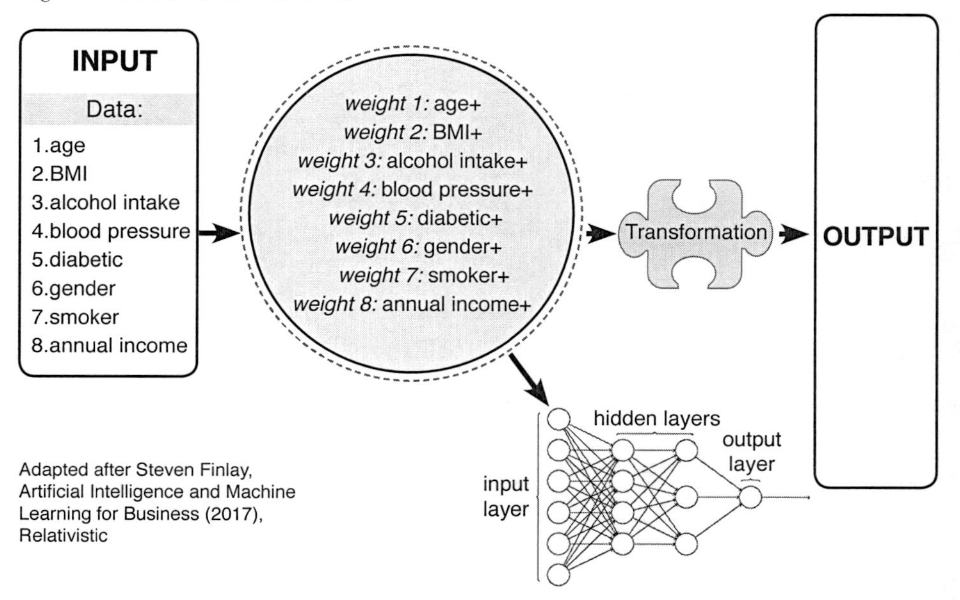

In 1973, Hinton first introduced a system with capabilities similar to those afforded by the fact that synapses change on multiple timescales. Yet, it was 43 years later in 2016 that he published a major paper on this area.[40] Andrew Ng, one of the leading names in deep learning, thinks that "deep learning is our best shot at progress towards real AI." Some of the most exciting and cutting-edge applications are currently being developed in the field of deep learning.

Challenges: This technique requires troves of data to train the algorithms, data which sometimes is not available or is very expensive. All of this requires a substantial amount of time. So it is costly in both data and time. In speech recognition, for instance, a multitude of dialects needs to be used for the algorithms to be thoroughly trained. Overfitting remains a fundamental challenge–an algorithm's efficiency is largely measured by how accurate it is when dealing with unseen data (e.g. a dialect which it has never encountered).

Applications: Deep learning has been remarkably successful in speech recognition and image processing, including video. You can find deep learning in Siri, Alexa, Cortana, and grammar correction tools like Grammarly, photo albums nicely organised in your iPhone, or in surveillance tools that analyse conversations to identify financial services regulatory breaches like insider trading.

Table 2.3 Common Challenges in Deep Learning Projects

COMMON CHALLENGES IN DEEP LEARNING PROJECTS	
HARD TO SOLVE WITH LOWER BUSINESS IMPACT	HARD TO SOLVE WITH HIGH BUSINESS IMPACT
• troubleshooting • testing	• effort estimation • resource limitations • privacy, safety, and regulatory requirements • black box - limited transparency • cultural differences • data science and business disconnect • unintentional feedback loops • putting AI models in production and scaling them
OTHER ISSUES WITH LIMITED IMPACT	
monitoring data tagging experimentation testing management	

Source: Arpteg,A., Brinne, B., Crnkovic-Friis,L., Bosch, J., Software Engineering Challenges of Deep Learning, 2018

What follows is a list of subsets to deep learning.

Convolutional Neural Networks (ConvNets or CNNs)

Definition: These are an approach to deep learning effective in areas such as image recognition and classification. The CNN is able to suggest relevant captions to photographs.

Challenges: While they are able to identify objects, they are not that accurate in identifying the correct location of each subitem in order to validate the whole object. CNNs struggle to differentiate between a croissant and a curled cat sleeping or a blueberry muffin and a chihuahua.

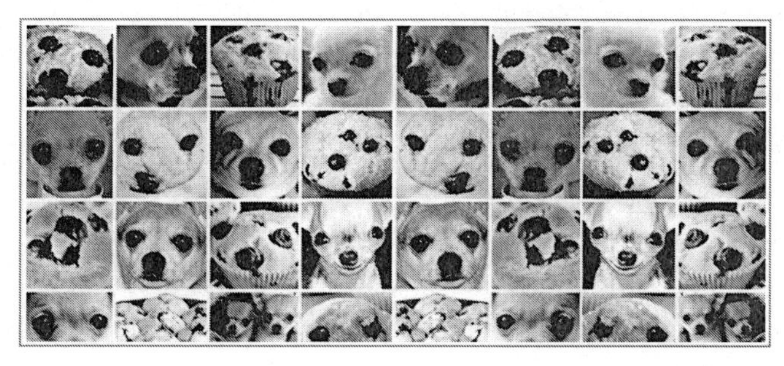

A clearer way to explain where CNNs struggle is that the agent/algorithm decides that it is a face as long as it can identify the defining elements (eyes, eyebrows, mouth, nose), even if the mouth is located in the forehead or the nose is in the ear. This causes issues in log-in authentication.

Applications: Highly effective in areas like image recognition and classification. ConvNets have been successful in identifying faces, objects, and traffic signs apart from powering vision in robots and self-driving cars. Lately, ConvNets have been effective in several Natural Language Processing tasks (such as sentence classification in simultaneous translations).

Capsule Neural Networks (CapsNet)

Definition: CapsNet are an advancement to address the spatial positioning and orientation with a superior dynamic routing mechanism (dynamic because the information to be routed is determined in real time).[41] In the example discussed earlier, the agent identifies that the mouth is not in the right place and concludes that it is not a face.

Challenges: Current implementations are much slower than other modern deep-learning models.[42] Time will show if capsule networks can be trained quickly and efficiently. In addition, we need to see if they work well on more difficult data sets and in different domains.[43]

Applications: Image recognition and computer vision with possible applications in risk management, mainly improved anti-fraud.

Recurrent Neural Networks (RNN)

Definition: RNNs are neural networks that make use of sequential information. They are called *recurrent* because they perform the same task for every element of a sequence, and the next output is based on previous computations. In other words, they have a *memory* of what has been done and base the next steps on that *memory*.

A good example of this is if you have an embedded sentence like "John didn't like Bill because he was rude to Mary." You process the beginning of the sentence, "John didn't like Bill," and then you use exactly the same knowledge to process "because Bill was rude to Mary." Ideally, you want to use the same neurons and the same connections and the same weights for the connections for this processing. In RNNs, if they're processing a sentence, they have to remember information about what has happened so far in the sentence, and all of that memory is in the activations in the hidden neurons.[44]

Challenges: RNNs are slow to train as they operate *sequentially* (e.g. translating words in a sentence one after the other) using *recurrence* (e.g. the output of each step feeds into the next). Longer sentences need more processing steps. However, in August 2018, Google launched a research paper "Moving Beyond Translation with the Universal Transformer" that aims to deal with this challenge. The breakthrough is that their model operates in parallel (translates each word simultaneously) while making use of a self-attention mechanism to incorporate context from words further away.[45]

Applications: RNNs have been used with great success in large-scale language understanding like translation. For smaller, more structured language understanding tasks, Neural GPU and Neural Turing Machine perform better. Uber uses it in financial forecasting for rolling forecasting.

Generative Adversarial Networks (GANs)

Definition: GANs are an emerging technique for both semi-supervised and unsupervised learning. They achieve this through implicitly modelling high-dimensional distributions of data.[46] In 2014, Ian Goodfellow and his colleagues at the University of Montreal published a groundbreaking paper that marks the beginning of GANs.[47] This is a two-layer neural network. They simultaneously trained them for a zero-sum contest: One neural network, called the *generator*, generates new data, while the other, the *discriminator*, evaluates it for authenticity. For example, the discriminator decides whether each instance of data it reviews belongs to the actual training dataset or not.
Challenges: Training and evaluating GANs from a statistical viewpoint (existence of divergences).
Applications: Images for the purposes of visualising new interiors of buildings, in e-commerce for visualising clothing items, or items for computer games; scenes; image enhancement with realistic textures, resulting in improved image quality.

Other Approaches to Machine Learning

There are a few additional approaches to machine learning, including semi-supervised learning, transfer learning, federated learning, and machine teaching.

Semi-supervised Learning(SSL)

Definition: This approach has been around since the 1960s, but it has become of significant importance due to the huge volume of unlabelled data coming from different sources like the Internet, genomic research, or text classification. It represents mixed learning algorithms, which typically add a small amount of labelled data into significantly larger unlabelled training datasets, for increased richness of insights. This is also expected to improve the accuracy of the outcome, task, or insight. There are several methods (self-training, co-training, transductive support vector machines, graph-methods, among others) that share the goal of taking advantage of the huge amount of (unlabelled) data to perform a classification.
Challenges: (1) Clustering has to work well for the unlabelled data and (2) identifying the correct amount of labelled data in order to be meaningful in relation to the amount of

unlabelled data. In 2018, a group of researchers proposed a new SSL algorithm to address these challenges.[48]

Applications: Customer engagement when customer segmentation needs a richer insight and a more accurate prediction.

Transfer Learning (TL)

Definition: This is a popular approach in deep learning that enables the transfer of knowledge (learning) from a model trained for a certain task to another model (which had access to vast resources to learn and to store knowledge), with a different yet somehow related task. It's like how we have transferring skills (mathematics) from one sector of specialism (portfolio investment) to another sector (computer science). Interest in transfer learning has doubled in the last 24 months, and advances have been made with notable progress in 2018 in natural language processing. Transfer learning is helpful when a given dataset has insufficiently labelled data to train an accurate model.[49] It is also useful to address the challenge of small datasets in smaller organisations like family offices, mid-size private equity houses, or credit unions. For further reading on the subsets of transfer learning, I recommend "Neural Transfer Learning for Natural Language Processing" (PhD thesis) by Sebastian Ruder.[50]

Challenges: Ensuring that the accuracy of the trained model is translated correctly to the new model,[51] how to measure knowledge gains, and transfer between more diverse tasks.[52]

Applications: Primarily used in computer vision (with successful applications) and natural language processing (still needs improvement and still in research stage) when the training of the first model requires vast resources (data and time). There are also applications in big data in the biology research field.[53]

Federated Learning (FL)

Definition: Federated learning (FL) is a collaborative machine-learning approach without centralising training data.[54] It was introduced by Google in 2017. In 2018, Intel advanced this approach with their work with the University of Pennsylvania to show the first proof-of-concept of FL to real world medical imaging[55]. They found that FL can train a deep learning model to 99% of the accuracy of the same model trained with traditional data-sharing method. The same approach can be envisaged for financial services. Federated learning (FL) uses the data but it does not move it from its source. Non-federated machine learning or traditional data-sharing method uses data that is physically moved into a pool (that is usually the cloud) from where it is then accessible. This is not always possible, sometimes because of (1) privacy concerns, (2) storage limitations and high-latency, meaning delayed access to data, (3) business competitive concerns, or (4) regulatory concerns. Federated learning aims to address these concerns by enabling machines (devices like mobile phones, iPads, etc.) to collaboratively learn while keeping the training data they learn from on their device and no individual updates are stored in the cloud. Some refer to federated learning as Decentralised AI or Confidential AI,[56] and they see it as a way to solve the data-privacy challenges, an area that will become even more important in financial services.

Challenges: This is a nascent type of learning, which still experiences a range of statistical and systems challenges.[57] Researchers are working hard to address them, and we can expect substantial improvements in the near future.

Applications: In general, any personalised applications that require user-specific data and need to respect privacy concerns and regulations; examples would be learning sentiment, semantic location, or activities of mobile phone users[58]; predicting health events like low blood sugar or heart attack risk[59] from wearable devices; or detecting burglaries within smart homes.[60]

Machine Teaching

Definition: In one of their 2017 research papers,[61] Microsoft postulated that they could make the process of teaching machines to program easy, fast, and universally accessible by increasing the number of individuals who can teach machines. This paper proposed that we should have a library of teaching programs to train a new system. Machine teaching becomes a new service and a new skill. In 2018, there are more than one million people who train AI.

Challenges: Existing deep-learning frameworks will need to be redesigned to adopt new types of more advanced computing hardware.

Applications: The most active infrastructure projects in this space include Google's TensorFlow's XLA, Intel's Nervana nGraph[62], and other dedicated projects like Mighty AI, Understand AI, Defined-Crowd, and Microwork. [63]

Genetic Algorithms

Definition: Sometimes referred to as evolutionary algorithms, genetic algorithms are search methods[64] that address the optimisation problem inspired by natural selection (Darwin's natural evolution theory) and have proven to be an exceptional search tool subject to the proper tuning of parameters.

Challenges: High sensitivity to the initial datasets–once it reads something it does not forget it irrespective of new information; inconsistency of outcomes that may vary even when run on the same datasets.

Applications: In 2019, IBM launched an innovative way to study neurodegenerative disorders with evolutionary algorithms by building a neuroscience model.[65] In automated trading, GAs are used to identify the best combination of parameters for choosing the best stocks and identifying trades, to match a trading strategy or recommend new trading strategies (buy, hold, and sell signals).[66]

OTHER MAIN SUBSETS OF AI

In addition to all of the specific machine-learning approaches, there are numerous advancements taking place in other subsets of artificial intelligence. Some of these include: natural language processing, computer vision, expert systems or rules engines, robotics, cognitive computing, and ambient computing.

Natural Language Processing (NLP)

Definition: NLP is multidisciplinary field and a subset of AI that seeks to enable machines to (1) understand the human way of using text and speech in a myriad of foreign languages and contexts, (2) learn this language, (3) use language in the right context, and (4) generate human-like writing.

Challenges: Understanding the context for either text or speech.

Applications: Digital assistants (Alexa, Siri, Google Home, Cortana) or conversational AI tools use NLP in customer service applications; sentiment analysis in portfolio management or customer feedback; surveillance of e-mails and phone calls for regulatory requirements; generate written investment reports; a person′s voice recognition in a group conference call; predictive replies in your e-mails, depending on the e-mail text, such as "Sorry to hear that. Thank you. I shall follow up." Or "Dear John," if John is who you are replying to. Google has been leading the research in NLP, and for further reading I recommend their language team's work on the Google AI website.[67]

Computer Vision (CV)

Definition: Computer vision is a multidisciplinary field and a subset of AI that seeks to enable devices, including physical robots, to understand their visual surroundings just as humans do. It does this through collecting and analysing imagery (photos and videos) and then acting.

Challenges: Human vision uses cognitive processes, geometric models, heuristics, goals and area plans, image transformation, classification, and so on. As discussed earlier under Convolutional Neural Networks (CNNs), spatial understanding can obstruct the accuracy of a visual model; the lighting of objects analysed may impact the quality of a machine's visual understanding, a challenge in autonomous cars. This is a field where advancements are happening quickly.[68]

Applications: Already used in fraud prevention and claims evaluation; iPhone's Face ID, which recognises its registered owner and automatically unlocks; payments in China are done by merely scanning a customer's face as a form of identification; visual immersion into investor's portfolio, allowing improved trading decisions; Know Your Customer (KYC) seamless on-boarding process for new customers; damage evaluation in claims insurance; and drones inspection of large areas of benefits insurance, investors, and other stakeholders.[69]

Expert Systems or Rules Engines

Rules engines are used to perform both forward and backward inference over a network of rules. Business Rules Management Systems (BRMSs) are descendants of expert systems from the 1980s, the golden age of this technology. It is still used in 2018, either on its own or in conjunction with other AI tools, primarily in solutions that require alignments with fixed regulations, such as audit, chatbots, or compliance policies.The rules are expressed in "if..., then ..." For example, "*if* the loan applicant doesn't have four months of credit history, *then* pull a report from the Credit Bureau."[70] Because the business rules change, it is important that the rules be correctly maintained. This is expensive and time consuming.

Robotics

Definition: Robotics is a subset of AI technologies that builds AI robots that interact with the physical world via sensors and computer vision and that display a level of understanding of this world and of movement (semi-autonomous or autonomous). Like with many other concepts, experts disagree about what defines a robot. However, one thing is clear–*robot* or *robo* is misused in relation to robo-investing or robo-advisors in our industry. An example of a robot might be Pepper, the programmable robot that you find in some bank branches following you around and offering to answer your questions. Depending on how they are programmed, AI robots can tackle problem solving, language understanding, and logical reasoning in how they interact with their outside world. Robots need to have vision so that they can perceive their surroundings. Robot vision is achieved as a combination of camera hardware and AI algorithms to allow the robot to process visual and sensory data from its surroundings, in order to control the motion of a robot by way of feedback from the robot's whereabouts and proximity to obstacles (objects, humans) as identified by its sensors.

Cognitive Computing

Definition: Cognitive computing is an amalgamated term defined as a "technology approach that enables humans to collaborate with machines." Some credit IBM's marketing teams with coining this term just to differentiate IBM's marketing messaging. IBM's Watson is an example of cognitive computing, which is described as ingesting vast amounts of data in the "cognitive stack" while creating connections within this information. Watson is not a single tool, but a mix of models and APIs.[71] Critics of the cognitive computing approach argue that this is data-structuring and that using "search" and "retrieve" functions does not amount to domain knowledge.[72] IBM's Watson program, which beat human contenders to win the Jeopardy challenge in 2011, was "largely based on an efficient scheme for organising, indexing, and retrieving large amounts of information gathered from various sources."[73]

Ambient Computing

Definition: A term purportedly first coined by Kevin Ashton in 2009,[74] it denotes the growing network of sensors and connected devices that surround us. Think of it as the digital ambiance that surrounds us just as our home ambiance does (interior decoration). For this digital ambiance to work it needs AI and by implication computing (a precondition of AI).
Challenges: Reliably integrating different devices, sources of data, and orchestrating complex event management; monitoring anomalies in such a complex ecosystem; security of anything from encryption, authentication of access, and recognition of different layers of permissions.
Applications: The digital ambience creates a network of suppliers of information and information flow (interior decoration would require colour coordination, the logic of feng shui, furnishing of our homes and offices, furniture suppliers, etc). In simple terms, the digital ambience around us creates an ecosystem. For instance, when our smart fridge is programmed to order more butter if it recognises that we are low, or when a Tesla car can be

controlled from our phone to come to our home, or when our iPhone identifies correctly the world surrounding it, including location, temperature, humidity, noise, radiation, and so on.

A range of business cases have been identified in logistics (inventory management, route optimisation), health and wellness (remote patient care), or manufacturing (connected machinery), with matching business value from basic efficiency and cost reduction to innovation, decision making, and customer engagement.[75]

EXAMPLES OF QUESTIONS FOR BOARDS TO ASK

	TOPIC	POSSIBLE QUESTIONS
1	Algorithms	1. What tasks do we want to achieve and what matching algorithms are we using? 2. How do we tackle data labelling? Do we always need it? 3. Have we considered reducing the costs by using algorithms that do not need labelled data? 4. What are the AI challenges that we have identified in production and implementation? What is the strategy to preempt them and solve them? 5. Have we allocated enough resources for the testing stage? How long is it expected to be?
2	Accountability	1. Do we have a case management log for AI implementation issues? 2. Out of all the AI projects that are active, how many of them are explainable? Can we explain these outcomes like the regulator would ask us? For instance, say if John has received the loan or not, or if Mary's loan is at a significantly higher interest rate? 3. Can we answer with 100 percent confidence the question "Why has the algorithm provided this outcome?" 4. Do we keep an audit trail of the work on AI projects and how we build the models, how we select the data, and how we address bias? 5. Who in our business is accountable for these models?

KEY POINTS TO REMEMBER

- AI consists of generated *predictions* based on mathematical models. Different algorithms are used to solve different problems, and sometimes they need to be mixed to find a richer solution to address a problem.
- There are different definitions of AI that correspond to two main approaches: Human-centred (based on observation and hypothesis about human behaviour) and Rationalist (based on a combination of mathematics and engineering).
- "Artificial intelligence (AI)" and "machine Learning (ML)" are two terms that are *incorrectly* used interchangeably. ML is a subset of AI, *but* AI ≠ ML, because AI is more than ML.
- The three types of machine learning are supervised, unsupervised, and reinforcement learning. Supervised learning (SL) is a subset of machine learning that works with labeled data and makes predictions based on a given set of examples, so there is a predetermined outcome. The goal of supervised learning is to optimise the mapping of a set of inputs to their correct output. Unsupervised learning (UL) works solely with unlabeled data and makes *predictions* on the patterns it finds, so there is no pre-determined outcome. This approach to learning is similar to how humans learn and solve problems. Reinforcement Learning (RL) is based on (1) trial-and-error and (2) reward to achieve progress with its task. The algorithm is not given a training data set, so it learns from its own experience by exploring what happens when it does something.
- Deep Learning (DL) is a subset of machine learning that attempts to simulate human neural networks in computer programmes, creating artificial neural networks.
- DL includes convolutional neural networks (ConvNets or CNNs), which are effective in image recognition and classification; capsule neural networks (CapsNet), which address spacial positioning and orientation; recurrent neural networks (RNNs), which make use of sequential information and have a memory of what has been done; and generative adversarial networks (GANs), which are a two-layer neural network in which one network generates new data (the generator) and the other evaluates the data for authenticity (the discriminator).
- Other approaches to machine learning include: semi-supervised learning, which adds a small amount of labelled data into larger unlabelled training datasets for richness of insights; transfer learning, which enables the transfer of knowledge from a model trained for a certain task to another model with a different yet related task; federated learning, which uses the data but does not move the data from its source; machine teaching; and genetic algorithms.
- Other subsets of AI include: natural language processing, computer vision, experts systems or rules engines, robotics, cognitive computing, and ambient computing.

CHAPTER THREE

Prerequisites for Artificial Intelligence

For as long as humans have existed, there has been data collection, storage, and analysis–either as sticks, stones, or bones–dating back to the Upper Paleolithic era. Early Egyptians used data in the form of hieroglyphics written on wood or papyrus to record the delivery of goods or calculate taxes.[76] Today we use algorithms to decode the meaning and draw insights from the data stored. In modern times the word "data" denotes images, numbers, and anything else in digital format that can be stored and transmitted electronically. Its modern-day meaning is traced back to eighteenth-century physicists and mathematicians like Newton, Gauss, and Lavoisier, who set the foundations of modern statistics methodology.[77]

According to International Data Corporation (IDC), the total data volume is expected to reach 35,000 exabytes in 2020. The world's total digital output was about 180 exabytes in 2006 and grew to about 1,800 exabytes in 2011–a ten-fold increase in five years (one exabyte = 1,000,000,000 gigabytes). Organisations actively using cloud, mobility, and big-data technologies are experiencing up to 53 percent higher revenue growth rates than those that have not invested in these technologies (Dell Global Technology Adoption Index GTAI 2015).[78] Google Chief Economist Hal Varian's insightful comment that "data is widely available, what is scarce is the ability to extract wisdom from it" provides us with a summary of the main challenges organisations have when dealing with a deluge of data coming their way every day.

To the financial services industry, January 2007 marked the beginning of a new era, and yet few really saw it coming. When Steve Jobs presented the world with the iPhone, he actually presented our industry with the beginning of an unprecedented *ocean* of data, which sometimes is referred to aptly as big data. Each data science problem has its own specific ecosystem to extract and interpret the data. Turing Award winner (1998) Jim Gray once made a prophetic observation: "We are seeing the evolution of two branches of each discipline: (1) data sourcing, collection, preparation, and analysis from many sources and experiments (2) data modelling for simulations of the working process using this data (e.g. simulations)."

As more datasets are being built, and new techniques will enable machine teaching as a job, there is a need for convergence–the overlapping of various datasets that can be trained at the same time. For instance, Andrej Karpathy, a Stanford PhD graduate and Director of AI at Tesla, has done essential work in computer vision and natural language processing for visual recognition with text captioning. This is why on the internet you can find live streaming with captions or photos with captions that describe that it is a white dog sitting next to a blue lake.

There are a number of hurdles to overcome with data sourcing and collection. In addition to the internal culture being resistant to changing the existing IT infrastructure, which seems the core challenge, there are a number of frequently observed challenges:[79]

1. data sharing and processing: lack of data governance frameworks
2. data preparation and analysis: lack of expertise to know what to remove unnecessary records and aggregate data, pivoting datasets, adding relevant information
3. data access: users of data analysis exceeding the IT professionals who provide them with the data analysis–one solution would be to train and allow analysts access to raw data
4. data collection: regulatory requirements (e.g. privacy concerns, GDPR)
5. data sourcing: correctly identifying bias and removing it
6. data infrastructure: lack of adequate investments in suitable infrastructure

This chapter will explain the different types of data, the infrastructure for storing data, how to source and collect the right data, the tools available for data analysis–including cloud computing–and the different types of computing hardware or chips. In order for an organisation to deploy AI, there are five main conditions to satisfy:

Table 3.1 Five Preconditions for AI Deployment

		WHERE IS IT DISCUSSED?
1	Data	This chapter - Chapter 3
2	Computing power	This chapter - Chapter 3
3	AI Algorithms	Chapter 2 - The Taxonomy of AI
4	Personnel	Chapter 7 - Strategic adoption of AI in Business
5	Business strategy - AI strategy	Chapter 7 - Strategic Adoption of AI in Business

In this chapter, I'll discuss data and computing power, the *back office* if you will of AI deployment, and I'll detail types of data and approaches to address the ever-growing need for computing power. As a board director, you may feel this chapter is a bit too technical, however, I recommend that you skim it first and use it as reference material. Chapter 2 (The Taxonomy of AI) and Chapter 3 (the current chapter) are heavy in technical references, but they are as essential to board decision making as reading the company's accounts. In the same way that not all directors are accountants but they need to understand the accounts, not all directors are computer scientists but they need to understand AI.

TYPES OF DATA

There are many different types of data, from structured to unstructured, internal and external, qualitative and quantitative, and even semi-structured, meta, alternative, and synthetic. It's helpful to understand the differences, because AI algorithms use data to perform different tasks. What follows is a structured list of types of data. They have many overlapping points, but it is useful to see the various kinds of data.

Big Data

The proliferation of smart devices and their interconnectivity has made the data size reach an unprecedented level. Following the "iPhone moment," researchers have worked hard and continued to do so in order to develop an entire ecosystem of hardware and software solutions for collecting, storing, processing, and analysing this *ocean* of data. Some of this ecosystem is still referred to as *Big Data*.

How does AI help with big data? Most of the data is unstructured (image, text, video, web, logs, etc.) and stored in raw format. AI algorithms perform well in extracting insights by clustering raw data in formats that are most suitable for the task at hand. The ability to extract value from big data offers corporate boards and decision makers an array of opportunities. The more executives see the rewards and competitive advantages in interpreting their organisation's data and identifying their data value strategy, the more investment in big-data initiatives will continue to increase. Big data offers enhanced insights essential to business decisions and process automation. It can be used for fraud detection, improving customer experience, risk management, compliance regulation reporting, and operational enhancement.

In 2001, Doug Laney of META Group (now owned by Gartner) wrote a forward-thinking piece on data management, as he noticed how business conditions were pushing traditional data-management principles to new meaning and formalised approaches.[80]

In his piece he introduced three dimensions of data (increasing volume, velocity, and variety), which have later been expanded by others:

1. volume – the ever-increasing scale of data
2. velocity – the analysis of movement of data
3. variety – different forms of data
4. veracity[81] – uncertainty of data based on its origin and how trustworthy it is
5. value[82] – ways to monetise the data to derive business value

Each of the different types of data discussed below are not mutually exclusive.

Structured Data

Either internal or external, structured data is typically easier to store and model, mainly because it assumes less work to clean it up and make it ready for modelling. This is the data that you'd find in anything from simple spreadsheets to comprehensive databases (for example, client relationship management, sales, trades, reconciliations, or financial accounts).

Around 70 to 80 percent of internal data is structured data. Specialists believe the benefits of structured data, which is more nuanced and detailed, outweigh the benefits of unstructured data and recommend a mix of structured and unstructured data to improve insights and predictions.

Unstructured Data

In simple terms, think of unstructured data as any piece of information that cannot be readily fit into a spreadsheet or a set template. For example, e-mails, video footage, images, client calls, recordings, Bloomberg chats, social media, or website interactions.

Each source of data might have a mix of video, text, and image, and with the right algorithm, you can have access to insights from this "messy" data pool. An estimated 80 percent of the world's data is unstructured. Unstructured data is growing at 15 times the rate of structured data. Customer analytics is one use case for unstructured data, a valuable source for marketing insights.

Semi-structured Data

As its name implies, it is a cross breed of structured and unstructured data. It does not conform to a predetermined approach found in structured data, and yet it contains some sort of identifiers like tags to separate it from the unstructured data. It is useful to consider when collecting data for modelling and algorithm training purposes.

Metadata

Typically each file contains data about when it was created, its purpose, the author, file-size length, and so on. These details form the metadata.[83] This enables large amounts of structured and unstructured data to be organised.

Internal Data

Internal structured data is the easiest to find, analyse, and model in the enterprise environment. Naturally, it is free for the business to access it. This is one of the key strategic assets any financial services organisation has–its current data or the data it is able to collect. This data must be maintained and kept secure, which are two expensive and necessary requirements. Internal data is a good starting point to extract valuable business insights, however, these insights have to be applied to the industry and macro external context, hence the need to weave in external data for richer insights.

External Data

External data comes in various forms and is anything that's outside internal data. It can be publicly available (see below for various free databases) or privately owned by a third party. External data brings nuance, context, and insight to internal data, however it does not always come free. Experian or London Stock Exchange data are two examples that have been widely used. We shall see below various ways to source external data.

Alternative Data

Alternative data comes from non-traditional data sources, such as satellite images or social media posts. It includes the following:

- Individual: conversation data, social media, news, reviews, personal data, electoral roll, web searches, photos, videos
- Business Process: transaction data, corporate, Government Agencies, internal security photos, and video
- Sensor: satellite, drones, geolocation, others

The *Financial Times* ran an article on how buy-side spending on alternative datasets has significantly accelerated over the last two years (~60% y/y) and is estimated to reach over $1 billion by 2020.[84] Uses of alternative data come in many forms. One example is how Schroders' data science unit used geo-location data to assess a pizza-chain expansion plan based on casual dining competition and foot traffic.[85] Industry leaders affirm this momentum in alternative data use. According to Chris Molumphy, CIO of Franklin Templeton fixed income, "The advanced pace of technological disruption is impacting the traditional investment landscape, providing new ways to identify and originate investment opportunities that generate value for investors."

AlternativeData.org is an example of a data aggregator that targets alternative data for hedge fund and long-only asset managers.

Qualitative and Quantitative Data

The researchers at Google, drawing on their recent success in advancements of neural networks, have started talking about *qualitative data* versus *quantitative data*. Volume does not necessarily mean quality. Building quality datasets to train algorithms is one of the most expensive and taxing preconditions of AI. Quality data is hard to achieve and it is very expensive, requiring extensive work. Currently, large datasets are mainly being used but some have free access for researchers:

- ImageNet, an image database pioneered at Stanford University in 2009, contains over 14 million images as of Aug 2018.[86]
- Microsoft and Google have introduced several additional high-profile datasets.
- The Canadian Institute for Cybersecurity specialises in cybersecurity datasets.
- The Canadian Institute for Advanced Research has built image datasets that have attracted a lot of attention: CIFAR-10 and CIFAR-100.
- Proprietary Dataset: TwentyBN, a German-Canadian company, is a computer vision company that built advanced machine-learning algorithms that understand video. They provide free access for academic research purposes to over 220,000 videos clips of humans performing.
- Government datasets provided for free or paid access.

In the mid-2000s, research suggested that the more data the better and that AI models need troves of data to produce more accurate results:

- In 2009, the Google research paper "The Unreasonable Effectiveness of Data" concluded that using large unstructured sets of data implied an improvement in the accuracy and quality of insight in translation tasks.[87]

- In 2011, the Microsoft research paper "Mitigating the Paucity-of-Data Problem: Exploring the Effect of Training Corpus Size on Classier Performance for Natural Language Processing" demonstrated how an algorithm significantly improved with more data.
- In 2012 Anad Rajaranman of Walmart Labs reached a similar conclusion at the end of a competition that aimed to build the best recommendation algorithm.[88]

However, recent developments in machine learning indicate that the size of datasets might not be a stumbling block in achieving accuracy of the outcome.

Synthetic Data

Synthetic data is data that has been algorithmically created, therefore it is not generated by real-world events but it mimics them. This type of data was first used in the 90s, but in recent years its effectiveness in essential business cases made it grow in importance, with key uses in machine-learning cases. In financial services its uses are:

1. Fraud protection, as it enables scenario planning and so improved ability to predict and anticipate new threats
2. Agent-based modelling (ABM) or scenario planning where fully synthetic data is used and the entire dataset is created. GANs (Generative Adversarial Networks) are algorithms that are able to create synthetic data. Partially synthetic is where sensitive data is replaced with synthetic data. For this reason some refer to this type of data as a subset of anonymisation of data.[89]
3. Addressing privacy issues by (1) replicating important statistical properties of real data without exposing it, or (2) generating datasets where real data does not exist.

DATA INFRASTRUCTURE

Requirements for storing and analysing data are business sensitive and form key imperatives for improved and seamless operations and client service delivery. Established technology companies like Google, Facebook, Amazon, and AirBnB are data-driven companies with a business model reliant on deployment of AI at scale in different jurisdictions. Their teams have identified solutions that work and continue to improve how they operate.

There seems to be a consensus that modern data infrastructure suitable to support AI would include the following: data lakes, data warehouse, cloud-based storage systems, internal enterprise servers.

Data Lake

This centralised repository allows storage of structured and unstructured data at any scale. Data can be as-is, without having to first structure the data, and it can run different types of analytics–from dashboards and visualisations to big-data processing, real-time analytics, and machine learning to guide better decisions.[90] The most important characteristic of the data lake is incorporating a machine learning framework like Apache Spark.[91]

Data Warehouse

A data warehouse is a federated repository for data that is aggregated from disparate sources within a business. This approach has proven highly efficient with structured data, such as finance function, which provides greater executive insights into corporate performance. They are expensive to scale down and give good performance within complex or unstructured data environments.

Cloud-based Storage Systems

There are free open-source softwares for most of the big data tasks, including storage. Storage systems are designed to run on cheap, off-the-shelf hardware and open-source software to store and analyse the data. However, this may not be a suitable solution if there is not enough in-house expertise and for data-safety requirements. There are paid-for solutions provided by Google, Microsoft, Amazon, and a host of other enterprise-grade providers.

Internal Enterprise Servers

In small organisations, using their own servers is a cost-effective solution if there is no need to store large amounts of complex data. In financial services this storage solution is slowly becoming obsolete due to the large volume of data. The industry standards set expectations on regulated organisations to retain client data for an extended period of time–in some cases even seven years or more.

DATA SOURCING AND COLLECTION

One of the main considerations in data sourcing is to identify and acquire appropriate datasets. The acquisition cost is an essential consideration. It involves the actual price paid to acquire a dataset as well as the opportunity costs of analysing data that might not be used (put in production). However, the largest datasets are privately owned by various organisations, and access to them usually costs hundreds of thousands of dollars in the early stages of a prototype development. In order to avoid some of these costs, in some cases public data may be a good starting point. There are a number of open datasets for AI, which are relevant to many sectors, but we will focus on those relevant to finance and economics.[92]

Table 3.2 Different Types of Databases

OPEN DATASETS RELEVANT TO FINANCIAL SERVICES	
Quandl	economic and financial data useful for stock price predictions and other economic indicators
World Bank Open Data	covers global population and economic indicators
IMF Data	international finances, debt rates, foreign exchange reserves, commodities prices, and investments
Financial Times Market Data	real-time information on financial markets
Google Trends	internet search of global activity and trending news stories
American Economic Association	United States macroeconomic data
UK Open Data	data published by central government, local authorities, and public bodies from business to justice, health, and education
Twitter	a useful granular tool for analysis of social media
OpenCorporates	the largest database of companies
Microsoft Azure Data Markets	extensive datasets across a multitude of sectors
PAID	
FitBit	150 billion hours worth of heart rate data–the largest dataset in history (also knows people's age, sex, location, height, weight, activity level, and sleep pattern) and in combination with the heart data is considered a goldmine on human health
DATA-AS-A-SERVICE	
Oracle	access to three trillion dollars in consumer transaction data; profiles of over two billion global consumers; tracks over 900 million users' web activity; world leader.
Data brokers that focus on credit reporting and risk assessment	
Experian Equifax Transunion	
Data brokers that focus more on target marketing and advertising	
Quantium Acxiom (lists over 700 million profiles) Epsilon	

Financial services organisations' data science teams need to know where they can access open-source datasets. This will enable them to build the digital environments they need to test ideas and simulate the real environment where they might be deployed, such as customer engagement. Training algorithms in these simulated environments would enable an organisation to decide how best to fine-tune those tools, or perhaps decide to buy rather than build. Below are a number of free databases based on the AI applications they need to train on.[93]

Table 3.3 List of Free Databases

SCOPE	EXAMPLES
Geospatial Data	OpenStreetMap; Landsat8; NEXRAD; NASA Exoplanet Archive
Networks and Graphs	Amazon co-purchasing, gephi.org, Amazon reviews
Speech Recognition	CHIME; TIMIT; VoxForge
Natural Language Processing (NLP)	Text Classification Datasets; WikiText; Billion Words
Computer Vision	MNIST; ImageNet; Visual Genome; archive.org
Recommendations and Rankings	Netflix Challenge, Movie Lens; eBay Market Data Insights
NLP Text Processing	Google Scholar, Microsoft Marco
NLP Voice Processing	TED-Lium (1495 TED talks audio recording)
Machine Learning	Machine Learning Dataset Repository (mldata.org); data.world

Data Collection Companies: Diffbot, Data Sift, Crowdflower, Crowd AI, Data Logue, ParselHub.

Sources of administration data publicly released by governmental institutions might come with a cost. For instance, David Chie, president of the San Francisco Board of Supervisors, believes that companies that use the data should pay 1 to 3 percent of transaction costs, and he is pushing to turn this into a law.[94] Sourcing data is one of the hurdles start-ups have to accessing enough relevant data to train properly. There are a number of companies that explore the intersection of Blockchain, AI, and data marketplace to create equal access to relevant and valuable datasets.[95]

Aggregation of anonymised and pseudonymised data may prove to address some of the privacy concerns. Open-banking is a key *data event* in financial services conferred by legislation that designs unprecedented access to customer data. While it opens many *data doors*, they need to be aligned to the regulatory frameworks set out, for instance, by the California Consumer Protection Act 2018 or General Data Protection Regulation (GDPR), which stipulates data minimisation as a core data protection principle under Art. 5, and breach of this principle results in 4 percent of global turnover or €20 million, whichever is the greater.

The organisations that have espoused the value of data aggregation must not rethink their approach, as financial services companies can only process customers' personal data necessary to achieve the purpose for which it was originally collected. This gives customers more transparency and control over their personal data. Accountability is much greater and it falls on financial services providers to respect the control of personal data to be in the hands of the customer.

Table 3.4 Companies Using Blockchain to Democratise Access to Datasets Needed to Train AI Algorithms

COMPANY	INVESTORS	SCOPE
Enigma	$25M in VC investments $45M in token sale	Catalyst allows companies and organizations to contribute data that users can subscribe to and consume via smart contracts, which tracks data, so the data cannot be distributed or resold further on, which is the current state of things.
Numerai	$7.5M in VC investments	Is a long/short equity hedge fund that has built a meta model crowdsourcing machine learning probes to a community of data scientists and then combines them in what becomes their trading strategy. They have their own token, Numeraire, which is used by data scientists to bet on how effective their solution is. Earnings of the fund as a result of this strategy are then tracked and data scientists remunerated.
Datum	$7.2M in token sale	Over 90,000+ people use Datum app to monetise their consumer data with pre-screened partners.
NuCypher	$5M in VC investments	A secure data-sharing platform for enterprises by encrypting private data and bringing this data to enterprise models
Ocean Protocol	$10.4 M in token sale	A global data exchange, where data are bought and sold securely
Computable Labs	undisclosed funding	A new platform designed to deliver a new Internet infrastructure to democratise access to data by providing a registry as a central point for the buying and selling of data.
Synapse AI	undisclosed funding	A market place for data and talent (researchers, trainers, processors, and contractors)

Without a clear focus on the business strategy and understanding of their own data strategy value drivers and capabilities versus the competition, organisations might find themselves wasting a lot of energy collecting and mining data for insights. Robbie Allen, Founder of Automated Insights, an enterprise natural language processing & generation platform, discusses some of the challenges of data collection: "We find out they have no idea how the data is stored, that it's distributed across a number of systems, and the technical talent required to get the data out properly is booked on other projects for months. That's how a three-month project easily blows out to six months or more."[96]

DATA PREPARATION AND STORAGE

Data preparation is the process of gathering, combining, structuring, and organising data so it can be analysed. Companies that are the most successful at adopting new AI techniques have synthesised core business functionality in the form of well-documented internal Application Programming Interfaces (APIs) and interfaces that make cross-department and cross-team technical collaboration easy. Many also deploy business intelligence (BI), analytics dashboards, and visualisation tools to help non-technical leaders and employees engage with important insights.

Data preparation is the essential step to ensure that the correct data is read to be put in production. According to Gartner, poor data quality accounts for 20 percent of business process costs. Data preparation can be partially outsourced to third-party providers, and indeed data preparation-as-a-service might become a high margin business.[97] In 2017, less than 1 percent of available data was analysed (Oracle), and 93 percent of execs believe they are losing revenue as a result of not fully leveraging the information they collect.

Open-source First Data Strategies

These strategies use open-source data-processing engines (running under Apache Software Foundation licence, a US non-profit), which offer free datasets and are open to anyone who needs them, and have proponents in data-driven companies like Google, Facebook, and Netflix, however a careful choice of an open-source data-processing engine is dependent on each business's needs.

- Apache Hadoop is an open-source platform of software utilities used for storage and processing for structured, semi-structured, and unstructured data using the

MapReduce programming model. Hadoop, named after a toy elephant, is a platform that emerged from one of Google's internal projects. It started as an internal tool for Google's File System and was discussed in a paper published in 2003,[98] and over the years it has evolved into a key platform with uses beyond search as originally designed. Hadoop is considered a breakthrough in big-data processing not only because it is fast but also because it dramatically reduced the cost of analysing and accessing vast deposits of data.[99]

- Apache Spark is a unified analytics engine for large-scale data processing[100] that uses machine-leaning techniques and runs on Hadoop, Apache Mesos, Kubernetes, standalone, or in the cloud. It can access diverse data sources and powers a range of libraries, allowing for ease in combining them.[101] It's main scope is speed of processing, providing a competitive advantage for its users. It was originally incubated and developed at the University of California and was released in 2014.

- Apache Storm is an open-source data-processing engine that focuses on extremely low latency (processing a large volume of data with minimal delay). Storm has flexibility with many programming languages and is used in real-time analytics, online machine learning, and continuous computation.

- Apache Presto was developed as a project at Facebook in 2012 and is an open-source, distributed SQL query engine, designed for fast analytic queries against data of any size. It supports both non-relational and relational sources and can query data where it is stored, without needing to move data into a separate analytics system. Query execution runs in parallel over a pure memory-based architecture, with most results returning in seconds. You'll find it used by many well-known companies like Facebook, Airbnb, Netflix, and Nasdaq.[102]

- Impala was developed by Cloudera and it is similar in scope to Apache Presto. Cloudera has joined forces with Hortonworks and announced in October 2018 that they will merge in a deal valued at $5.2 billion, which they hope will give them a more powerful platform across multiple clouds on premise and edge computing but also with customers and industries. The combined platform will have more than 2,500 customers and about $500 million in cash. The new entity will be able to compete in streaming, IOT, data management, data warehousing, machine learning, and hybrid cloud markets, according to Hortonworks CEO, Rob Bearden.[103]

The stored data volume is a growing cost and in some cases can increase exponentially because of the volume and velocity of data. Therefore, Data Strategy design needs to deal with and anticipate two critical issues (1) data retention (what to keep for storage) and (2) data disposal (what to remove permanently). Machine-learning tools can be used for different tasks needed as part of data strategy, for instance in data preparation.

DATA ANALYSIS AND MODELLING

There is an ample source of open-source technology available for data analysis and modelling. Data analysis sits on three main pillars:

1. data preparation – refining raw data by collecting, cleaning, and consolidating it.
2. data modelling – defining and analysing data requirements to address business' needs.
3. data analysis – extracting information from the data and creating reports.

MapReduce is a common open-source method used for data analysis. Oracle, Cloudera, Microsoft HDInsight, Amazon Web Services, IBM, and Google also provide paid commercial software to turn data into insights. They all have their own competitive offering, but when selecting the most suitable data analytics tool, the reader would be well advised to have a thorough understanding of each provider's strengths and weaknesses. Most of these commercial offerings still use open-source Hadoop (see earlier in this chapter) and build on it for varied analysis tools. For instance, Goldman Sachs Asset Management (GSAM) is focused on creating data-driven investment models that can objectively evaluate public companies globally through fundamentally based and economically motivated investment themes. They credit their infrastructure's ability to capture and process data quickly and have given them new levers to capture investment themes.[104]

Table 3.45

Goldman Sachs Asset Management Quantitative Investment Strategies Approach to Identifying Investment Opportunities - Big Data Investment Approach	
Momentum	use machine learning to identify the connections between companies based on industry sentiment, stock movements, and correlations in economic factors
Value	analyse a large universe of industry-specific data that extends beyond a company's financial statements to determine its "intrinsic" value
Profitability	evaluate a company's web traffic patterns to identify businesses that are gaining e-commerce market share in real time.

Source: GSAM Perspectives, The Role of Big Data in Investing 2016, and interview with Osman Ali, Takashi Suwabe, and Dennis Walsh, Portfolio Manager sGSAM Quantitative Investment Strategies,[105]

Data Training-as-a-Service

This service is an emerging lucrative space because organisations' internal AI teams need it since AI experts are in short supply and expensive. When building internal AI capabilities, certain work can be done with small data science teams and outsourcing data training and preparation.[106] See the examples below:

Table 3.6

DATA PREPARATION COMPANIES	USE CASES FOR DATA PREPARATION COMPANIES
Trifacta	risk compliance and security; fraud detection; customer behaviour and segmentation; personalisation
Paxata	insight discovery; risk compliance; customer behaviour analytics
Alation	searchable catalogue for insights and decision support
Tamr	uses machine learning to clean up and integrate data in business context
Alteryx	enables users to blend data, build analytics, and share business insights
Pentaho	open source-based platform that offers reporting, analysis, dashboard, and data-mining solutions for big data deployments
DefinedCrowd	also offers specialist training data for financial services
Syncsort	allows organizations to collect, integrate, sort, and distribute their data

Feature Engineering is key to data modelling and how successful a machine-learning project will be. In plain English, it means that the team (1) understands the task at hand and identifies the strengths and limitations of the algorithm selected to be used, and (2) experiments and tests the algorithm/model to see what works and what does not. Feature Engineering is "often data type specific and application dependent."[107] This means that each model has its own peculiarities and generalisation does not work. This is an essential detail to keep in mind when selecting AI systems.

Deep industry knowledge is key in data modelling, and that is one of the much-needed skills for successful deployment of any AI models. For instance, Automated Insights, which enables a company to turn quantitative data, such as sports scores and earnings reports, into computer-generated articles, requires domain expertise to produce the best results. The hard part is determining what is interesting and worth talking about in a story. Nara Logics, a company based in Boston, Massachusetts, aims to bridge the deep sector knowledge with the help of "*synaptic intelligence,*" which aims to replicate rather than imitate the brain circuits' mathematical logic. They draw primarily on neuroscience and modern engineering. Their platform "learns from messy, siloed data" and unites it for "better customer recommendations and employee decision support."[108]

DATA STRATEGY IMPLEMENTATION PRINCIPLES

A traditional company can become a data-driven business by following certain principles. In financial services not all companies are ready for artificial intelligence. Enterprise-scale AI is not standalone plug-and-play technology. The quality of your information architecture, IT infrastructure, and your organisation's design and culture are critical foundations for any AI initiative.

Table 3.7

	DATA STRATEGY IMPLEMENTATION PRINCIPLES
1	Start with business goals and then fine-tune your supporting technology.
2	Define your enterprise standards for data quality.
3	Build bridges between business and data science teams.
4	Define and implement data standards.
5	Improve your organisation's data literacy.

Principle 1: Start with business goals and then chose the technology.

Technology follows business strategy–this is a core concept. This means that we first need to decide what we want to achieve as a business (e.g. increase profitability by accessing new markets, by increasing efficiency, by attracting 10 million clients in two years), and then select the technology that can get us there.

Therefore, start with the business objective you want to solve, define it clearly, state its measures of success, and then allow your data science teams the space to evaluate options and scenarios, test hypothesis, refine the outcome, and also asses the technologies that are there in place and provide an audit to the board. They will need time to test. At this stage it is easier to find out which technologies work and which fail. It is essential to have proper budgeting.

Principle 2: Define your enterprise standards for data quality.

Data quality is the key stumbling point in any AI initiative. Quality of data means using the right data for the business objective that you want to solve. It is a misconception that you need a large volume of data for AI, so any data you have access to should be thrown in. A large volume of data is expensive to prepare and store, and it can also introduce unnecessary "noise" in the insights. Quality of data is a nuanced exercise, because this is the entry point where bias can be removed or introduced.

Therefore it is essential that the teams working at this stage have diverse backgrounds, genders, and views. They are the insurance policy against bias, and they can also build stronger defence mechanisms to fight points where datasets are skewed. The truth is a product of perceptions. Data governance is the measure that ensures data quality, and it needs to be a protocol at the foundation of corporate governance for any enterprise. It falls within the oversight function of boards. The implications of not doing it right create reputational risks and litigations, which can damage or simply wipe out a business.

Principle 3: Build bridges between business and data science teams.

The disconnect between these teams can create confusion and bottlenecks in AI initiatives. It is rare to see data engineers physically sitting next to business teams. They should learn about each other's work, challenges, and what their perceived successes look like.

A data science team's success does not always translate to business success. For computer scientists this disconnect begins as early as their university degrees. It is only recently that these degrees include electives on how businesses operate and how a computer science degree can deliver value in a business setting.

Principle 4: Define and implement data standards.

Access to data shouldn't be reserved for only a few people in the company. Disseminating access to the data across the firm is essential, however, it has to be done based on strict access protocol. Data veracity is essential. In order to establish its veracity, data needs to be consolidated, cleaned, and accessed with dedicated tools that are known and understood enterprise-wide.

Data access protocols need to be part of any instruction sessions for new joiners and regular updates for current employees. These protocols are living protocols that also need to meticulously document employees' confusion when using them and how their concerns have been addressed.

Principle 5: Improve your organisation's data literacy.

Data literacy is essential across the organisation–from top management down to the most junior level. Data literacy can be compared to Microsoft Office literacy. Everyone is expected to know how to use Excel or type a letter in Word, irrespective of whether they are a board director or the receptionist. Dedicated training needs to be embedded in the regular compulsory training to ensure that everyone understands the data-usage protocols. For decision makers digital and data literacy translates into improved financial results.

CLOUD TECHNOLOGIES AND AI-AS-A-SERVICE

The National Institute of Standards and Technology (NIST), a technology agency that is part of the US Department of Commerce, works with industries to develop guidance standards and measurements. According to NIST, cloud computing is "a model for enabling ubiquitous, convenient, on-demand network access to a shared pool of configurable computing resources (e.g., networks, servers, storage, applications and services) that can be rapidly provisioned and released with minimal management effort or service provider interaction." Cloud computing is linked to cloud storage and is used to work on and complete specified projects; it works on task with data that is only stored in the cloud. It requires higher processing power than cloud storage, because unlike cloud storage–which is only about data storage and access–cloud computing gives the user the ability to remotely work with and transfer the data stored.

Data storage will become of core importance to financial services. Relying on regular cloud providers, which seem unlikely to want to be regulated for storing the financial services' data, opens a range of strategic concerns. The large majority of organisations typically have a multi-cloud strategy for cost management but also for risk management reasons. And yet, I would argue, that one underlying risk might not be properly managed. Financial activity (spending, saving, and investments) provides an intimate and accurate insight into individuals' behaviours and behaviour triggers and is possibly 10 times more accurate than social media activity (the number of likes of cats, dogs, or baby pictures can't tell that much). Institutions need to ensure strict Service Level Agreements (SLAs) with cloud providers.

It is essential that the industry, its regulators, and boards take a closer look at where and how the data is stored and accessed, while rushing to innovate. It is useless if one spends time securing the windows, only to leave the main door wide open.

In recent years cloud providers have changed their offering to embed AI technology. Worldwide spending on cognitive and AI systems totalled $12 billion in 2017, an increase of 59.1 percent over 2016, according to International Data Corporation (IDC), a global provider of market intelligence.[109] Such estimations are based on the proliferation of the need for these systems and data input. Mission-critical applications stored on the cloud raise a range of risk issues typical for outsourcing infrastructure to a third party. These are risks that need to be mitigated and balanced with the advantages of the cloud. Infrastructure costs are high, but cloud computing brings a much-needed cost reduction and efficiency in software usage, deployment, and operational efficiency. Cloud computing is a highly lucrative industry. There is a tight battle among the enterprise cloud service providers (currently these are mainly Google, Microsoft, and Amazon), which are consistently upgrading their services with an ever-more sophisticated AI layer to complement their cloud-computing offering. The industry is expected to reach $513 billion by 2022.

Clearly organisations are interested in building their cloud strategy, and according to a survey they indicated that AI is of particular interest.[110] This has brought about a new business model: "AI-as-a-Service" or "on-demand-AI." This is a third-party offering of artificial intelligence outsourcing. It allows individuals and enterprises to experiment with AI for various purposes without a large initial investment and with lower risk. Experimentation can allow the sampling of multiple clouds to test different types of algorithms. This has been considered the next frontier, and providers are looking to provide AI applications delivered through the cloud. There are some notable developments:

Table 3.8 Providers of AI applications Delivered Through the Cloud

NVIDIA	In 2017 they announced a new cloud service supporting major frameworks such as Tensorflow, Torch, Coffee, and CNTK. Processors are hosted in the cloud as needed.
Amazon Web Services (AWS)	Established a third-party developers' program for Alexa and use a range of ML frameworks available (machine learning, deep learning AMIs, Apache MXNEt and Tensorflow on AWS); they offer a range of APIs like Comprehend, Polly, Rekgnition, Translate and Transcribe.
Google Cloud	Launched Cloud Natural language API, which can be used for text comprehension in several languages; also launched Cloud Video Intelligence to search and discover video content and which quickly annotates videos stored on Google Cloud; they partner with organisations including Cloudera and Talend to deliver an enhanced and competitive offering for enterprise needs, including competitive pricing and no pre-paid lock-in.

IBM Cloud	Provides a range of APIs under Watson Machine Learning and allowed third-party developers to build cognitive applications leveraging Watson platform; partnered with Twilio, a cloud communication platform used by over 1 million developers
Bonsai	Democratising development of AI applications and systems for non-AI programmers
Next IT	Alme Platform for virtual healthcare providers
Salesforce	They launched myEinstein, an APIs that offered customised CRM tool for enterprise needs without needing data scientist.
Oracle AI Platform Cloud Service	Oracle embeds ready-to-use AI and machine-learning capabilities across Oracle's SaaS, PaaS, and IOT services, including cognitive AI, analytics, data services, IT management, and security operations.
Intel	In 2017 they announced that Intel Nervana DevCloud would be available to 200,000 developers, researchers, academics, and startups to learn, sandbox, and accelerate development of AI solutions with free compute cloud access powered by Intel® Xeon® Scalable processors.
Microsoft Azure	Offers APIs with strict regulatory requirements for financial services; solutions for regulatory compliance; embodied Cortana, their conversational AI. Microsoft, under its cloud offering, has developed its Azure Sphere product, an intelligent edge solution to power and protect connected micro controller unit-powered devices and they are the ideal fit for various markets like aerospace, defense, automotive, medical, etc. Microsoft approach to next wave of computing–intelligent edge and intelligent cloud. One of the advantages is in increased data security, when you take the power of the cloud to the device–aka the edge or the next wave of computing - provides the ability to reason, make decisions, and act in real time in areas where there is no connectivity or limited access (source: https://blogs.microsoft.com/blog/2018/07/23/the-next-wave-of-computing-is-the-intelligent-edge-and-intelligent-cloud/).

Currently the AI-as-a-Service product falls into three main categories:

1. *Specific Applications:* digital assistants, computer vision, natural language processing, emotion detection, knowledge mapping.
2. *On-demand Service:* allows developers to build machine-learning models that improve with more data.
3. *Fully-managed services:* used when the enterprise lacks ML expertise and works on drag-and-drop development tools that are pre-built.

Benefits of AI-as-a-Service:[111]

1. *Advanced AI Infrastructure* – multiple servers and fast GPUs that run workload in parallel.

2. Low Cost – eliminates expensive up-front payment for hardware.
3. Scalability – easy to start with a pilot project that can be moved into production and scale up.
4. Usability – AI capabilities are made available without requiring expertise.

Limits to making AI-as-a-Service relevant:

1. *Specific Sector Knowledge:* Each major cloud provider has recognised this limitation and has implemented financial services-specific teams, to enable them to provide a sector-specific offering and understanding.
2. *AI-Enabling Infrastructure:* In the past decade we have seen an unprecedented level of complexity and interconnectivity of channels or networks used to convey, source, and collect various types of data. These channels are expected to converge in order to adapt to enterprises' needs like enhanced customer service, anticipating cybersecurity threats, or the convergence of different systems applications. Managing and understanding this complexity has brought into discussion the work of Claude Shannon, a former Bell Telephone Laboratories mathematician and MIT professor, who in 1948 published a research paper considered the foundation of the field of Information Theory.[112] His work looks into how communications channels have a theoretical maximum limit of information transfer. This is referred to as Shannon Limit or Shannon Capacity. There are important discussions about how the industry is fast approaching the limit beyond which the data would degrade because there wouldn't be enough channel capacity to manage it safely. In order to avoid reaching this limit, specialists are looking into new ways to provide network architecture to manage the scalability problem presented by the traditional server-centric applications[113] by designing either where data resides (cloud and end-point) or selecting what data is relevant to a user in a given moment (spatial division multiplexing, which is all about integration and referred to as information networking[114]).

COMPUTATIONAL RESOURCES/HARDWARE

The creation of a high volume of data is driving rapid technology progress in data storage. In 1980, the world's first gigabyte-capacity hard drive, the IBM 3380, was the size of a refrigerator, weighed 550 pounds (about 250 kilograms), and had a price tag of $40,000[115]

Nearly 40 years later, you can pick up a gigabyte of storage on a postage-sized memory card for a few sterling pounds[116] or the value of a lunch meal. The AI-enabling infrastructure–the computing power behind the digital world–was identified as an investment theme in 2016 in one of Bank of America Merrill Lynch's reports, which estimates that the accelerated computing processor chip market across cloud, supercomputers, and enterprise applications has the potential to grow 10 times to over $10 billion by 2020.[117] The pace of change is sometimes referred to as "cambrian explosion."

Moore's Law and the Advent of the Microchip

According to Gordon E. Moore, co-founder of Intel, "The microchip is made of silicon, or sand–a natural resource that is in great abundance and has virtually no monetary value. Yet the combination of a few grains of this sand and the infinite inventiveness of the human mind has led to the creation of a machine that will both create trillions of dollars of added wealth for the inhabitants of the earth in the next century and will do so with incomprehensibly vast savings in physical labour and natural resources."[118]

Silicon, the principal ingredient in beach sand, is a natural semiconductor and the most abundant element on Earth except for oxygen.[119] It is one of the most important semiconductors today and the building block of most electronic devices. Semiconductors are used to fabricate chips for every electronic device, including computers, cell phones, iPods, BlackBerry devices, and GPSs. Each chip may contain a million devices that perform different functions. A combination of science and engineering enables us to turn sand into different types of microchips, such as the following:

- Logic chips that perform the computation inside most commercial platforms
- Memory chips that store information and are called RAM (random access memory)
- Application-specific integrated circuits that are special-purposes chips used on cars or appliances[120]

The data flow between a processor, memory, and peripherals is managed by an integrated circuit (IC), referred to as a "chipset." Typically a chipset is bespoke to the same type of microprocessors. Electronic circuits are fixed on a small flat piece–referred to as "chip"–made of a semiconductor material, usually silicon. A chip is a complex device that forms the brain of every computing device and cloud platform. The integration of large numbers of tiny transistors into a small chip results in circuits that are smaller, cheaper, and faster in magnitude than those constructed of discrete electronic components.[121] The speed to switch on/off has increased as the size of transistors has become smaller. The larger the number of transistors, the smaller the cost and size of devices, and the higher their functionality has taken off in the past decades.

But let's return to Gordon E. Moore for a moment. One cannot write about chips, computation power, and silicon technologies without mentioning Moore's famous quote: "The number of transistors incorporated in a chip will approximately double every 24 months." He noted this back in April 1965. Termed a "law" years later by Caltech professor Carver Mead, Moore's Law went on to become a self-fulfilling prophecy, which anticipated competition and mega trends in the silicon or semiconductors technology industry for over half a century and the state of the *digital revolution*.

The chip industry, also referred to as the silicon industry, has enjoyed some 50 years of phenomenal growth and stellar financial results on delivering ever-growing computational power. They would release a new chipset every 18 months that was twice as powerful as the previous one. This model has been working brilliantly, and why change something that works? However, transistors can only get so small. This poses a physical limitation to Moore's law. Intel is already working on a 5 nanometre sized node, which is very close to the limit, while the current industry standard node size is 14 nanometres and this might cause a stop in the growth of computational power.[122]

In the search for more computing power, chip makers have started working with software architects to find pockets of power by optimising the chip's performance. The competition is fierce as this industry is now in search of innovative ways to organise computation. In Chapter 4 we will discuss in detail the forces and strategies at play in the industry. Why is this important to know? Because this industry's competition will influence how AI algorithms will develop in the next two-year horizon. This is a valuable indicator to help decision makers buy technology that will be relevant for longer.

Computing Hardware

There are two main types of computing hardware:

1. Traditional or von Neumann architecture typically uses:

- CPUs *(Central Processing Units)*
- GPUs *(Graphical Processing Units)*
- ASICs *(Application Specific Integrated Circuits)*
- FPGAs (*Field-Programable Gate Arrays)*
- IPUs *(Intelligent Processing Units)*
- AI chips *(Artificial Intelligence-ready chips)*

John von Neumann was a Hungarian American born in Budapest (Neumann Janos) to an affluent Jewish family. He was a child prodigy who could at age six divide two eight-digit numbers in his head and converse in Ancient Greek. He turned into an accomplished mathematician, computer scientist, and polymath who contributed prolific research in mathematics, nuclear weapons, and computing.

He is considered a founding figure in computing and also worked with Alan Turing on the philosophy of artificial intelligence. He was married to Klara von Neumann, also a Hungarian-born mathematician. Together they immigrated to the United States in the early 1940s. Klara was an accomplished computer scientist in her own right and became the head of the Statistical Computing Group at Princeton University and then went on to leave her mark by programming two of the early computers: MANIAC I and ENIAC. Klara and John's aggregated legacy in computing is known as *von Neumann architecture.*

2. Emerging architecture, which emulates the human brain, typically uses:
- Neuromorphic
- HIVE (Hierarchical Identify Verify Exploit)
- Photonics-based approach

CPUs (Central Processing Units)

Moore's Law, while it was prescient, also created self-fulfilling prophecies for the industry to continue placing central processors on one single chip. This architecture is referred to as von Neumann architecture or general purpose architecture. It assumes a central processor unit (CPU) responsible for executing a computer program. CPU was first developed by Intel in the 1970s. It has one *core,* which means that it can perform one operation at once. In 2001, IBM released the first *dual core* processor, which was able to focus on two tasks at the same time. This meant that more CPUs have been crammed into one processor. Some modern computers have at least 40 CPUs. CPUs are suitable for tasks like parsing a complex set of commands, however they might not cope with machine-learning algorithms' needs for expandable processing power. CPU models are fast approaching their capacity for improved computing power.

GPUs (Graphical Processing Units)

Graphical Processing Units (GPUs) have been around in gaming applications since the 1970s and by 2018 they are the standard in personal computers. While CPUs perform *one complex* task at a time, GPUs can do *many simple tasks* in parallel. Researchers have identified that the GPUs' architecture is more suitable for machine-learning algorithms.[123] GPUs also have high computational precision across the board at all times. This is a downside, because it is taxing on memory bandwidth and data throughput flow. GPUs have often recorded 10 times the speed in ML-training models compared to CPUs. NVIDIA is the most well-known manufacturer of GPUs and they have recorded a steady growth reflected in a growing stock price. NVIDIA's platform, Cuda, is effectively the only useable one for machine-learning applications today.[124] This is why GPUs are a promising chip for improving chipsets to handle machine learning and particularly deep-learning algorithms, which are hungry for vast memory,[125] data, and computing power. GPUs are in high demand right now for AI, gaming applications, and even cryptocurrency mining.

ASICs (Application Specific Integrated Circuits)

An ASIC is a processor designed for a specific type of task. While GPUs can perform many simple tasks at the same time with high computational precision across them, this is taxing on memory bandwidth and data throughput, because this high precision is not always needed. ASICs narrow down to executing a specific function or type of functions. Google has developed its own ASIC processor called Tensor Processing Unit (TPU), designed solely for machine learning with Tensorflow (their own deep-learning framework). TPUs are the only ASICs widely available and they are also available to the general public.[126] Their TPU architecture is 15 to 30 times faster than the best CPUs and GPUs for programming neural nets,[127] and it iterates faster, is more energy efficient, is suitable for running AI on IOT sensors (known as "on the edge"), has improved memory bandwidth, and enables cloud infrastructure as a service. They have rolled it into Google Cloud Platform and offer a great interface by allowing companies to build models on Google's competing chips like Intel's Skylake or GPUs like NVIDIA's Volta and then move the project over to Google's TPU.[128] In the same spirit of deep competition and to attract talent, they offered a cluster of 1,000 Cloud TPUs for researchers working on open machine-learning research.[129]

In January 2018, Samsung announced that they began building ASIC chips for mining cryptocurrency. Given the range of improvements ASIC chips promise, we expect to see traditional and new chip manufacturers focused on function-designed chips, because it enables them to squeeze complex computation into a chip.

Cambricon is a Chinese chip manufacturer that develops chips across different classes. Their chips are in the Huawei's Kirin NPU IP, they are used in data centres through a PCIe card, and their latest MLUv01 architecture is considered NVIDIA's major ASIC competitor.[130]

FPGAs (Field-Programable Gate Arrays)

FPGAs are bespoke chips. They are designed to be one-time programable by the final user to their desired application after manufacturing, therefore they allow greater flexibility. This feature distinguishes FPGAs from Application Specific Integrated Circuits (ASICs), which are custom manufactured for specific design tasks. ASICs and FPGAs have different value propositions, which must be carefully evaluated,[131] and they have gone mainstream as they are competitive on performance and costs.[132]

Cisco and Microsoft are two examples of companies using FPGAs in their hardware. Microsoft uses FPGAs to power its search engine, Bing. Amazon Web Services, Amazon's cloud offering, has moved to FPGAs in the cloud. Large chip manufacturers have started bringing hybrid hardware to the market, which is required during different stages of running an AI inference engine for a range of destinations like drones, factory robots, and automobiles, where there are environment-specific requirements like latency, data processing speed, and low power consumptions. NVIDIA brought hybrid hardware platforms to the market with an ARM/GPU combo in NVIDIA's Jetson and DrivePX2. Xilinx and Intel offer high performance systems off the chips that bring together FPGAs and ARM.[133]

IPUs (Intelligent Processing Units)

This chipset is designed for a deep-learning environment and was developed by the British company Graphcore. They successfully raised $50 million in December 2017, as they promised substantial development in handling advanced AI algorithms up to 10 times faster than the current processors. In 2019, they raised an additional $300 million from Microsoft and BMW among others, which valued the company at $1.7 billion. The founders of Graphcore had previously launched a semiconductor company, which they sold to NVIDIA for $435 million in 2011.

The IPU's first chip, Colossus, has more than 1,000 processors that communicate with each other to share the complex workload required for machine learning. Poplar-heir proprietary software- moves data across its chip more efficiently with less wasted processing power and uses all processors sequentially.

AI-ready Chips

Apple has announced that they are working on their dedicated chip, which powers iPhoneX, for processing images and speech. Their chip powers the algorithms that recognise your face to unlock your phone and it would also enable other features. Experts think that their "neural engine" could become central to future iPhone models and that leading mobile tech companies would follow Apple's example and develop their own neural engines.[134]

Huawei's Kirin 970 features a dedicated Neural Processing Unit (NPU), which combines cloud AI with native AI processing and has an eight-core CPU and a new generation 12-core GPU that delivers 12 times the performance and 50 times the efficiency and can process 2,000 images per minute. They also offer the technology to developers and partners. In April 2017 Qualcomm announced that they had explored a strategic partnership with Facebook to help them develop technologies useful to machine learning on their Snapdragon processor. In April 2018, Facebook, in a heel turn, announced that they are developing their own AI chip. The reasons might be to lower their dependence on chipmakers like Intel and Qualcomm Inc.

In March 2018 Intel launched Movidius Myriad X, which is a product launched with Microsoft to enable AI inference at the edge (edge computing) for IOT systems and claims to be the "industry's first system-on-chip with a dedicated neural computing engine for hardware acceleration of deep learning inference on the edge"[135] and it will be offered to Microsoft users worldwide. In March 2019, Intel announced that together with the US Department of Energy they are building Aurora, the world's first exascale computer, the most powerful computer in the US and possibly the world. Aurora, an estimated $500 million project, will be completed in 2021 and will be available for academic research in the science, defence, and energy sectors. Aurora is so fast that it can calculate in one second what the entire world population can calculate in four years.

In February 2019 Tesla announced that they had built their own AI chip to build their own ecosystem to deliver what they want to deliver for the Tesla experience. In doing so, they will stop using NVIDIA's AI chip. Alibaba announced that it is working on its own AI chip, called Ali-NPU (which stands for neural processing unit). In September 2019, Intel launches its own AI chip Nervana NNP that addresses the pressing need for deep-learning training and inference at an exceptionally large scale.[136] Intel's work in financial services is particularity notable as they try to address with their technology specific industry requirements.

Neuromorphic Approach

Developed in the late 1980s, this is a computer architecture that is similar to biological brains and is an emerging chipset approach. Neuromorphic architecture includes functional units, which replicate neurons, axons, synapses, and dendrites, so it matches the biological brain's flexibility. In recent years it has been applied in new and different ways, including neurotrophic chips, which are superior in processing capacity than traditional microprocessors. Intel Labs developed the Loihi chip, which includes 130,000 neurons optimised for spiking neural networks.

HIVE (Hierarchical Identify Verify Exploit) Approach

A new paradigm of computing, this non-von Neumann processor could improve computing speed by 1,000 times. HIVE approach is designed to perform different processes on different areas of memory. It has been supported by DARPA, the American military's Defense Advanced Research Projects Agency funding, and it involves Intel and Qualcomm alongside academic centers.

Photonics-based Approach

In February 2019, Lightmatter, a Boston, Massachusetts, start-up, announced that they developed a scalable platform for high throughput–high efficiency artificial intelligence computing[137]–which they will be developing with the backing of Google Ventures, the Alphabet's VC arm. According to the Lightmatter founders, there are two critical problems in data centres: (1) *throughput* or how many operations can be performed per second, and (2) *efficiency* or how many operations can be performed per second per watt of power. Their systems will be capable of greater than 10 times existing solutions. Lightmatter has already produced an early chip that contains more than a billion transistors.

EXAMPLES OF QUESTIONS FOR BOARDS TO ASK

	AREA	QUESTIONS
1	Business problems to solve	1. To what extent does the problem to be solved relate to the overall strategy of the business? 2. Does it form part of the business strategy sanctioned by the board or executive committee? 3. What are the business gains if the problems are solved (e.g. cost reduction, improved client services, etc.)? 4. Who is the lead on this big data initiative and what's their driver? What's their input: to hinder or help? 5. Which department or process would benefit from applying AI to operational decisions? 6. Why wasn't this problem solved before? Did it need AI or not?
2	Cloud storage	1. Who has access to our data? How is it stored? What is our data storage strategy? 2. Where is our data stored? Where are the servers? 3. Are our Service Level Agreements (SLAs) clear enough that the cloud provider does not have access to our data? How do we make sure they do not have access? 4. Is our cloud provider(s) supportive of our disaster recovery program? 5. What challenges have been identified by our Chief Risk Officer? 6. Can we at least consider building our own industry-wide cloud (applicable perhaps at country level, a discussion for Government and supervisory authorities) and assess the costs and implications?
3	Data Sourcing and Training	1. Do we have quality training data? If not, how do we plan to get it? 2. Do we have an evaluation procedure built into our application development process to assess what works best? 3. If we use pre-packaged AI components, do we have a clear plan for how we will go from using those components to having a meaningful application output?

	AREA	QUESTIONS
4	Data	1. How well do the data scientists who build the mathematical models understand financial services with each sector's peculiarities (e.g. asset management)? 2. How well understood are the regulatory requirements and common-sense requirements specific for a client product design, for instance? 3. How accurate and complete is the dataset they will use to train the model on? How different is the training dataset to the same dataset? 4. Are there any "Lorenz's butterfly" effects expected on any datasets? 5. Models usually imply assumptions and simplifications of real-life deployment. What are the assumptions and simplifications? 6. The outcome of the models might get very close to human evaluation. How close? Why so? 7. Can we make our Terms and Conditions more simple to read and understand for our customers? 8. Are our terms of service written in an easy-to-understand language to explain how data will be used and how they can take back their consent to use data? 9. How easy have we made it to deny consent to use data? If it is not one click away, can we make it that easy?
5	Data lakes	1. Who will the data lake serve and for which business use cases? 2. Will the primary users be data analysts, data scientists, data engineers, or a combination of these roles? Do the users have the right skillsets? 3. What discovery and exploration tools can help unlock the value of the data lake quickly and continuously? 4. What ultimately defines the return on investment in the data lake? What are the short-term and long-term goals?
6	Chipsets	1. Do we know our end application requirements? 2. What kind of ML development stage do we need the chip for? 3. What type of software tools are available from the chip providers for training or inference environment (Tensorflow, PyTorch, etc.)? 4. Which section of Neural Network Exchange Formats standards are relevant to us? 5. What's the memory (on-chip and off-chip) bandwidth? 6. What's the throughput and latency? How do they correlate? Which is higher and how does that sit with our application requirements?

Sources in the table are as follows: Data Sourcing and Training[138], Chipsets[139], Data Lakes[140]

KEY POINTS TO REMEMBER

- There are many different types of data, from structured to unstructured, internal and external, qualitative and quantitative, and even semi-structured, meta, alternative, and synthetic.
- The ability to extract value from big data offers corporate boards and decision makers an array of opportunities. Big data offers enhanced insights essential to business decisions and process automation. It can be used for fraud detection, improving customer experience, risk management, compliance regulation reporting, and operational enhancement.
- Either internal or external, structured data is typically easier to store and model, mainly because it assumes less work to clean it up and make it ready for modelling. This is the data that you'd find in anything from simple spreadsheets to comprehensive databases.
- Think of unstructured data as any piece of information that cannot be readily fit into a spreadsheet or a set template. For example, e-mails, video footage, images, client calls.
- Internal data is one of the key strategic assets any financial services organisation has–its current data or the data it is able to collect.
- Data infrastructure includes data lakes, data warehouse, cloud-based storage systems, and internal enterprise servers.
- Data preparation is the process of gathering, combining, structuring, and organising data. Data analysis sits on three main pillars: data preparation, data modelling, and data analysis.
- The quality of your information architecture, IT infrastructure, and your organisation's design and culture are critical foundations for any AI initiative.
- "AI-as-a-Service" is a third-party offering of artificial intelligence outsourcing that allows individuals and enterprises to experiment with AI for various purposes without a large initial investment and with lower risk.
- The two main types of computing hardware are von Neumann architecture and emerging architecture, which emulates the human brain.
- CPUs have one core and can perform one complex task at once; GPUs can perform many simple tasks in parallel; ASICs are designed for a specific type of task; FPGAs are designed to be one-time programable by the final user to their desired application; IPUs are designed for deep learning; and AI-ready chips are designed specifically to handle very large datasets with speed and efficiency for deep learning models.

CHAPTER FOUR

The Current State of AI—Evolution, Growth, and Investing

In this chapter we'll examine the current AI landscape and investments in AI start-ups as a strategic entry point for financial services incumbents to access this technology. We'll also look at the maturity cycles, evolution waves, due diligence considerations for transactions in AI, and strategic investments in semiconductors and enterprise AI. Financial technology, also known as "fintech," has had tremendous growth in the past decade, and artificial intelligence can help these companies with their data storage, mobile services, fraud detection, cybersecurity, and more. In order to succeed and grow, today's financial institutions must incorporate cutting-edge technology into their business strategies. It's also critical that companies who choose to buy rather than build AI tools for their business understand the difference between real AI products and pseudo-AI products that market their software as AI when, in fact, it does not use adaptive intelligence to optimise tasks.

I believe that boards need to have a framework in order to evaluate how they should approach investing in AI. Naturally, their corporate venture capital (CVC) teams will do the investing, however, boards needs to familiarise themselves with a few key concepts:

- Technology maturity cycles
- Waves of AI evolution
- Why they should invest in AI
- How to invest in due diligence considerations or use AI to invest
- Venture Capital (VC) and Corporate Venture Capital (CVC) investments
- AI companies landscape
- Academia in AI research in financial services
- Government's role in innovation and how industry-led support can help

Since the 1950s, AI technologies have been through a range of ups and downs, mostly determined by the level of funding of what used to be regarded as just an academic curiosity. Since the 1950s there have been two *AI winters*—two periods of stagnation when the funding dried up—one from 1974 to 1980 and the other from 1987 to 1993. It's hard to see how we would encounter another AI winter, given the current state of AI developments. Many financial companies are accelerating change by developing partnerships with AI companies or creating their own innovation labs to develop innovative AI solutions. We'll look at some of the key players in artificial intelligence, fintech banking and finance, and AI-as-a-service providers who are developing their own computing infrastructure.

The majority of breakthrough developments have occurred roughly in the last five years. The speed of change has been without precedent and has been driven by advances in deep learning. "The revolution in deep learning has been very profound. It definitely surprised me even though I was sitting right here," Sergey Brin, Co-Founder of Google, stated with total candour.

In the past five years, I believe one name remains at the core of the fundamental shift in the funding and narrative of AI technologies: DeepMind. DeepMind is a London-based AI company that created (1) a neural network that learns how to play video games in a similar manner to humans and (2) a neural network that may be able to access an external memory, resulting in a computer that mimics the short-term memory of the human brain. When Google acquired them in 2014, for just $525 million, they "acquired" some of the best minds in neural networks as well as DeepMind technology, which super-charged many of Google's processes, including their search engine, which has propelled Google into this formidable presence in the search space and across many other verticals. In 2017, Google's CEO announced that they "are now an AI first company," having invested systematically in machine learning for more than a decade, which was the secret sauce to dominating the space. DeepMind was the inflection point of their investments.

And then an even bigger move came from DeepMind: In 2016, DeepMind built its AlphaGo program, which defeated the world champion at a professional GO game. The players of the Game of Go are highly regarded in China–where this strategy board game was invented more than 2,500 years ago–and rightly so, because it is overly demanding on human thinking capabilities. So, when DeepMind's AlphaGo turned out to be better than the world champion (who was reduced to tears when the game ended and he lost), it must have struck a chord with people who understand the deep meanings of the GO games. Many call that very moment "China's AI moment." In 2017, China announced their 2020 AI strategy. Since then, they have taken this objective very seriously. AI is being deployed at a scale and a speed that are not only amazing but also simply worrying. Many countries and businesses have published their AI strategy since then. In February 2019, the US came to the fore and presented to the world its own strategy. On the backdrop of the media frenzy covering AI, countries rushing to adopt AI strategy, and the European Union desperately trying to create some ethical order in this AI rush, many people started to put forth the idea of democratisation of AI as a remedy to the abuses of this technology by bad players. Elon Musk was among the first to promote this idea. Google invested in an open-source machine-learning toolkit called TensorFlow. Their purpose was to attract AI talent. They also wanted to control new AI applications, so they created a custom hardware designed to execute tensor calculations called TPUs (Tensor Processing Units). While democratising access to AI, they also amassed allegedly the world's largest collection of data.

People worry that surveillance capitalism is creeping into our democratic societies, placing at risk one of the most treasured values we have in our western world: democracy. *The Age of Surveillance Capitalism* is a book by Shoshana Zuboff, a professor at Harvard Business School, which talks about the challenges of humanity, the new economic order, and surveillance with our consent. It is recommended reading, especially for those people who have the power to make investment decisions in AI technologies and influence how technology should protect democracy and privacy–not destroy them. Wherever the money is invested, there will be growth. And yes, it might make more money for investors, but what is the actual cost to humanity, to our society, to our climate? Just as climate risks are investment risks for long-term investors, AI risks are investment risks. As the previous AI winters demonstrated, where there's no money, there's no AI development.

TECHNOLOGY MATURITY CYCLES

Carlota Perez's research examines major technological revolutions since the industrial revolution and how they were impacted by capital markets. Her book *Technological Revolutions and Financial Capital* is one of the books that I recommend as essential reading. Venture capital investors often refer to her work. Unsurprisingly, Anthemis, a leading VC firm at the centre of a vibrant ecosystem of start-ups and financial institutions and headquartered in London, launched Anthemis Institute led by Carlota Perez as academic-in-residence. Essentially, her research has found that for every technological revolution there are two phases:

1. *installation phase* when infrastructure is built for the new technology to come to the market (e.g. network infrastructure for the internet)
2. *deployment phase* when such technology is widely adopted (e.g. iPhones)

Between the two phases there is a *turning point,* which is always marked by a financial crash and then recovery, because "nothing important happens without crashes." A slightly more detailed cycle might be **invention → commercialisation → proliferation → consolidation → stagnation → disruption.**[141]

As discussed earlier in the book, there is a range of AI technologies, and each has their own maturity cycle driven by data, algorithms, and computing power. It's important to understand each AI technology separately and examine where they are in their surge cycle. Chapter 2 presents a taxonomy of main AI technologies and an easy-to-follow summary of their characteristics.

AI EVOLUTION WAVES

Defense Advanced Research Projects Agency (DARPA) is an agency of the United States Department of Defense responsible for the development of emerging technologies for use by the military. DARPA made key contributions to the establishment of the early internet and many other breakthrough technological innovations and is a leader in the field of artificial intelligence.

According to DARPA there are three waves of AI:

(1) Past: described as *symbolic AI,* with manually input knowledge, which had strong reasoning but poor learning capabilities. Algorithms were devised according to exact rules and the knowledge posted by their creators and other specialists. This approach is the core of chess-playing computers. Our smart phone apps also operate on this approach and Google Maps travel optimisation AI employs this approach.

(2) Present: described as *statistical learning,* which has strong learning capabilities but less strong reasoning. Experts do not build exact rules like in Wave 1; instead they develop statistical models for certain types of problems like credit decisioning, and then they train these models on various data samples to fine tune them for precision and efficiency. Statistical models are successful at understanding the world around them–they can learn and adapt to different situations if trained accordingly. Their solutions are calibrated to work well enough most of the time. Artificial Neural Networks, the technique underpinning deep learning, is representative of this approach and it is discussed extensively in Chapter 2. Facial recognition, speech transcription and translation, and autonomous cars and aerial drones are examples of ANN successes.

Yet, experts do not know how and why they work this well, hence many refer to them as black boxes. This opacity raises many ethical implications and governance issues, which are discussed in Chapter 5: Corporate Governance and AI Adoption.

(3) Future: described as contextual adaptation, which is considered the future of AI, which will be at the intersection of the two previous waves. It is expected that AI systems themselves will build models that will discover by themselves the logical rules that drive their decision-making frameworks. These AI systems will be able to train themselves just as Alpha Go did when it played more than a million games against itself to find the logical rules of the game. These systems will also be able to take information from all areas and adapt it to the context where the decision making is needed. For this phase to develop, according to DARPA, "there is a lot of work to be done to be able to build these systems."[142]

Table 4.1 AI Evolution Waves

AI WAVES	REASONING	LEARNING	HOW IT WORKS	WHAT IT LACKS
AI 1.0 : Symbolic AI	strong	weak	using predetermined exact rules	flexibility of learning
AI 2.0 : Statistical	weak	strong	using statistical models bespoke to certain problems	explainability of how and why they work
AI 3.0 : Contextual adaptation	strong	strong	AI systems autonomously build models and train themselves	explainability

Adapted from the Three waves of AI by DARPA summary

INVESTING IN AI FOR PUBLIC RELATIONS PURPOSES

There are a few founders of fintech AI start-ups who would recall how their strategic partnership with banks (legacy institutions) took a surprising turn. Here's what upsets start-up founders: Some legacy institutions would invest in AI start-ups not for long-term lasting change but for short-term PR purposes, to portray themselves as forward thinkers embracing the AI disruption with both arms. PR stunts would push an AI start-up's name in the media, but that's where everything ends.

In recent years, many organisations have randomly invested in fintech, blockchain, and AI as a measure of their more or less articulated commitment to "do something about" these emerging technologies. Unsurprisingly, a vague commitment renders a shallow outcome. When fintech first emerged, "doing something about fintech" was interpreted by some leaderships as a clarion call to invest "a little bit" in fintech start-ups in order to prove that something was done and then use those start-ups as marketing show-ponies. No one wins. I was talking about AI adoption with a large European bank, and they proudly told me that they are heavily into AI because (1) they have invested in 10 fintech companies and (2) they have built a data lake for their customer service bot. On the latter, we have seen earlier in the book that building a data lake is not necessarily the measure of a visionary take on AI adoption. On the former, mere investments in fintech start-ups mean nothing to business transformation with AI, unless these investments are chosen strategically to address ways of reimagining the business model. It's like buying ingredients to cook dinner at home, putting them in the fridge, and then going out to eat.

VENTURE CAPITAL (VC) AND CORPORATE VENTURE CAPITAL (CVC)

Looking at the global market trends in VC in 2019, Fred Wilson, managing partner of Union Square Ventures, a New York City-based VC fund, concludes that "there is a lot of money in the venture capital ecosystem right now." Crunchbase reports that the global VC market reached about $330 billion of the global deal volume. They also document an emerging trend in VC for some companies to raise $100 million or more in a single round of funding.[143] PWC/CB Insights Money Tree Report[144] echoes the same findings, showing that VC markets have moved to larger and larger deals while the number of transactions has declined comparable to 2017. In the period analysed, the total capital invested almost tripled while the number of transactions has declined. Fred Wilson thinks this is unsustainable, an opinion that I echo. While it is fashionable to invest in more mature companies, the up-and-coming AI companies are usually small start-ups. Traditional financial services companies have a small window of opportunity to invest in these AI start-ups at low valuations.

Table 4.2 Venture Capital Deal Size (2013 - 2018)

YEAR	INVESTMENTS USD BILLION	DEALS	YEAR	INVESTMENTS USD BILLION	DEALS
2013	36.4	5,176	2016	63.8	5,679
2014	60.5	5,998	2017	76.4	5,824
2015	78.1	6,098	2018	99.5	5,536

Source: PWC/CBInsights Money Tree report 2018

The report notes that AI-related companies raised $9.3 billion in 2018, representing a 72 percent increase from 2017. Fintech is also drawing more venture money. US fintech companies raised $11 billion last year, up 38 percent from 2017. The cryptocurrency firm Coinbase raised $300 million in October 2018, representing the largest fintech deal in the fourth quarter. In this environment, it is ever more essential for board of directors to set a business strategy that clearly embeds a sustained investment program in targeted AI technologies that will promote their future business growth.

As discussed below, CVC investment activities are in full swing. Global financial institutions like Goldman Sachs, Société Générale, and Mitsubishi UFJ Financial Group have rolled out clear investment strategies in emerging technologies. I am not at all surprised by the Mitsubishi UFJ initiative,[145] having had the opportunity to hear their President talking about their vision of the future in 2017 at FinSum in Tokyo.

DUE DILIGENCE IN AI TRANSACTIONS

Investing in AI companies is a pathway for financial services incumbents to buy their way into new technology. Such investments can be done either as pure investments and/or a doorway to access the technology with the view to deploy it in the business. Irrespective of the underlying reasons, I recommend that the board and executive teams have a framework to inform themselves about the particularities of due diligence in AI transactions. Covington & Burling LLP, a US law firm, drafted a useful yet non-exhaustive guide authored by Lee Tiedrich and Daniel Gurman. This guide rests on two core steps:[146]

Understanding the Transaction

For some, it might seem obvious that understanding the core asset of an AI transaction is key to seeing how it fits in the company strategy. In very simple terms, an AI company has a combination of the following core assets: (1) computing power, meaning semiconductors or chips, (2) software, meaning AI algorithms, (3) data, meaning data-training sets and their results/outputs, and (4) talent, meaning a team of talented data scientists. You need to be clear which one is the core asset to your transaction and how this core asset is relevant to your business. This enables you to focus on the due diligence specific to the type of asset.

Beware that there will be many so-called start-ups branded as AI companies. Some are merely a rebranding of an input data analytics tool or a spreadsheet with a fancy user interface. Others are simply the sum of freely available rudimentary algorithms. The *Financial Times*' analysis reveals that "investors should stop funding companies that simply tick AI on to their pitches" and instead "should focus on fundamental questions. What problem does this use of AI solve? How do you measure outcomes? How does the AI provider capture its share of value? These questions must be answered convincingly for an investment to succeed."[147] In 2019, David Kelnar, Partner and head of research at MMC Ventures, a London-based VC firm, did a study of all European companies self-described as AI and found that 40 percent of them are not AI. AI focused, AI driven, and similar descriptors may indicate that they plan to use machine learning in the future.[148]

Focused Due Diligence

Depending on the type of core asset an AI company has, the due diligence needs to focus on the following:

1. **Intellectual Property.** The core audit is to ascertain to what extent the software that implements the AI algorithms is derived from open-source or third-party software and who owns the output of AI software (details usually found in customer and other commercial agreements). Open-source audits are recommended together with the evaluation of the intellectual property protection of any design or additional developments of the software. Another consideration regards the data sources and the training data in order to identify what it is, how it has been used, and where it is sourced from. It should be noted that there have been litigations over web-scraping practices, so it is important to evaluate if there is any risk of litigation from the data sources. In this context it is worth mentioning that in January 2019 IBM launched a collection of nearly one million photos taken from Flickr, the photo-hosting site, and promoted it as a major step in the progress of reducing racial bias in facial recognition algorithms. IBM annotated each photo with details to include facial geometry and skin tone. Photographers, whose images were used, were disconcerted that their artistic work, which they had rights over, was used in this way without their consent. People whose faces were annotated this way were upset, too. After this information was leaked to the press, IBM walked it back and offered to remove the photos of anyone in this database who didn't consent to the use of their photos in this way.
2. **Data Privacy.** This regards the assessment of the compliance with the applicable privacy laws, and any other frameworks that govern data collection, processing, and usage. It includes personally identifiable information. In the absence of legal frameworks, a thorough application of common sense across the internal rules of the target company might be useful.

3. **Cybersecurity.** Cybersecurity experts usually work with legal teams to evaluate the legal implications of findings. The core concern is legal implications in the case of data breach and cost of remediation and reputational risk.
4. **Insurance Due Diligence.** "A cybersecurity incident often implicates a patchwork of different types of insurance policies, including cyberinsurance, general liability, technology errors and omissions, property, directors and officers, and crime policies," according to the authors of a Covington & Burling LLP article, which summarises and recommends an in-depth diligence of these policies to also include all limits and deductibles. This helps the potential buyer identify the level to which they are insured in the case of a cybersecurity breach.
5. **AI Governance.** Internal governance principles regulate how AI algorithms are designed, deployed, and how their outcome is used and if it is appropriate for the tasks intended. They also assess the legal risk of a particular solution. Governance of AI will only grow in importance and will continue to attract the attention of policy makers and regulators. At the point of investing it's essential to evaluate the level of awareness of this issue. It is investors' responsibility to invest in AI systems that promote fairness and are ethical. The Governance of AI chapter discusses in detail the range of topics executives need to be aware of.
6. **Product Liability.** There is a potential for product liability litigation, as with any other non-data-driven product. Monitoring the product development and audit trail of errors and incidents are key points in product liability due diligence. In addition, an audit of liability and risk management of contractual terms, including disclaimers and indemnification, is important.
7. **Surveillance.** This encompasses diligence of any lawful process requests and procedures in place to respond to government requests for surveillance that allow authorities to use surveillance laws to access digital information.
8. **Foreign Investment/National Security.** The Foreign Investment Risk Review Modernization Act came into force in August 2018 in the US and monitors foreign investments above 10 percent in certain businesses, such as technology companies, including robotics, autonomous systems, AI, and big-data companies. Similar legal requirements might already be in place in other jurisdictions, and they need to be thoroughly evaluated and the correct authorities need to be notified as required.
9. **Other Regulatory Requirements.** Extra due diligence is required to comply with regulations specific to financial services. Each jurisdiction has their own financial

services regulator and their requirements need to be closely followed. When in doubt, always seek extra guidance from the regulator and ensure compliance.

10. **Company's Executive Management.** Often investors find that their investments perform well but the companies they invested in display executive management failures, sometimes in the shape of cutting corners. Revolut is a London-based company hailed as a unicorn in fintech. In 2018 they raised yet another successful series at a valuation of more than $1 billion. In 2019, it transpired that there were regulatory issues (there are claims that management decided to switch off their money-laundering algorithms) and problems in their personnel recruitment process (people would be recruited only after they demonstrated that they signed up a few hundred new clients in a couple of weeks, a practice that was criticised as exploiting free labour).

AI COMPANIES' LANDSCAPE

In Feb 2019, CB Insights analysis identified the top 100 AI start-ups redefining industries. I met and spoke to a number of these companies. They deserve to be on this list. A total of 11 companies on the list are unicorns (private companies valued at more than $1 billion). In addition to the list of top 100, it is worth paying attention to those AI companies in Semiconductor, Enterprise Tech, Legal and Compliance, and Finance and Insurance as well as Media (fake news detection that is particularly essential in running quality sentiment analysis). CB Insights have proven to be pretty accurate with identifying the next generation of successful AI companies. For the latest updates always check out *www.cognitivefinance.ai*

Table 4.3 CB Insights' Next Generation of AI Companies

AI.Reverie	Enterprise Tech	Training Data	United States
DefinedCrowd	Enterprise Tech	Training Data	United States
Mighty AI	Enterprise Tech	Training Data	United States
Dataiku	Enterprise Tech	Data Management	United States
Machinify	Enterprise Tech	Data Management	United States
DataRobot	Enterprise Tech	Data Management	United States
Tamr	Enterprise Tech	Data Management	United States
H2O.ai	Enterprise Tech	Data Management	United States
Trifacta	Enterprise Tech	Data Management	United States

Dremio	Enterprise Tech	Data Management	United States
SigOpt	Enterprise Tech	Data Management	United States
mabl	Enterprise Tech	Software Development	United States
Applitools	Enterprise Tech	Software Development	United States
Demisto	Enterprise Tech	Cybersecurity	United States
Anodot	Enterprise Tech	Cybersecurity	Israel
Shape Security	Enterprise Tech	Cybersecurity	United States
Vectra Networks	Enterprise Tech	Cybersecurity	United States
Area 1 Security	Enterprise Tech	Cybersecurity	United States
Agari Data	Enterprise Tech	Cybersecurity	United States
Jask Labs	Enterprise Tech	Cybersecurity	United States
PerimeterX	Enterprise Tech	Cybersecurity	United States
BounceX	Enterprise Tech	Ads, Sales, & Marketing	United States
Unbabel	Enterprise Tech	Ads, Sales, & Marketing	United States
Gong	Enterprise Tech	Ads, Sales, & Marketing	United States
Gamalon	Enterprise Tech	Ads, Sales, & Marketing	United States
FullStory	Enterprise Tech	Ads, Sales, & Marketing	United States
UiPath	Enterprise Tech	Other: RPA	United States
Orbital Insight	Enterprise Tech	Other: Alternative Data	United States
Descartes Labs	Enterprise Tech	Other: Alternative Data	United States
Element AI	Enterprise Tech	Other	Canada
SparkCognition	Enterprise Tech	Other	United States
Prowler.io	Enterprise Tech	Other: RL Platform	United Kingdom
4Paradigm	Finance & Insurance	Anti-Fraud	China
BioCatch	Finance & Insurance	Anti-Fraud	Israel

DataVisor	Finance & Insurance	Anti-Fraud	United States
HyperScience	Finance & Insurance	Back Office Automation	United States
AppZen	Finance & Insurance	Auditing	United States
LawGeex	Legal, Compliance, & HR	Contract Review	Israel
Eigen Technologies	Legal, Compliance, & HR	Contract Review	United Kingdom
Onfido	Legal, Compliance, & HR	Onboarding & Compliance	United Kingdom
Textio	Legal, Compliance, & HR	Augmented Writing	United States
AI Foundation	Media	Fake News Detection	United States
New Knowledge	Media	Fake News Detection	United States
Habana Labs	Semiconductor	Data Centers	Israel
Graphcore	Semiconductor	Data Centers	United Kingdom
Cerebras Systems	Semiconductor	Data Centers	United States
Horizon Robotics	Semiconductor	Edge Devices	China
Thinci	Semiconductor	Edge Devices	United States
Syntiant	Semiconductor	Edge Devices	United States
Mythic	Semiconductor	Edge Devices	United States

THE AI FINTECH ECOSYSTEM

The Bank of England defines fintech "as technology-enabled financial innovation that could result in new business models, applications, processes, or products with an associated material effect on financial markets, institutions and the provision of financial services."[149] Fintech is building a transformative momentum in financial services. This began in 2007, with the launch of the iPhone, the world's first smart phone. No one could have predicted this change in 2007. In the same year, the CEO of Microsoft, Steve Ballmer, famously said, "There's no chance that the iPhone is going to get any significant market share. No chance." History tells us a different story. In fact, it was a historic moment. What followed was an unprecedented growth of smartphone penetration and affordability, an enormous volume of data at a high velocity, and the emergence of new technologies that enabled financial services institutions to capture and analyse large amounts of data. This has stimulated innovation, which has had a deep effect on how people access financial services, with 74 percent of the UK's adults using online banking, while the average bank branch received 104 visits per day in 2017, a 26 percent drop from 2012.[150] At this rate, in the coming years, fintech is expected to enable new product providers and new business models.

All financial services organisations are becoming fintech companies. Some are more technology driven, while others are slow to adopt technology. They have at their fingertips a vast amount of data and this is their common denominator. The key technologies that have been impacting financial services can be summarised as one of the following:

- *Big Data and Analytics* as the starting point for many business intelligence projects
- *Cloud* as a precondition to store the data and access it as is when needed for processing
- *Mobile Services* primarily used by mobile-only banks and more recently by incumbents, which are tapping into the data for expanding their data offering
- *Artificial Intelligence* to address many problems, such as:
  - biometrics for digital authentication
  - fraud detection and cybersecurity
  - digital assistants
- *Internet of Things* to enable a more nuanced and varied coverage, which has proved beneficial for insurance with the use of telematics

Basing their product delivery on these technologies, some fintech companies have built a full-stack offering and have progressed at a rapid pace. Such an example is WireCard AG, a 19-year-old payments and fraud prevention provider based in the Munich suburb of Ascheim, which has a market value of 21.1 billion euros ($24.1 billion), bigger than Germany's biggest bank, Deutsche Bank.[151] Adyen NV is another fintech payment company that registered a meteoric rise, in addition to Stripe, and the phenomenal growth of Chinese companies WePay and AliPay. For instance, AliPay counts about 100 million daily users and about 180 million transactions per day, and they account for just over 50 percent of the Chinese mobile payment market. Oaknorth, one of the fastest growing UK SME lenders, has gone from a loan book of £0 to a loan book of £1.5 billion in just 30 months.[152]

Fintech has arrived like a knight in shining armour expected to correct many of the mistakes the incumbents inflicted on their customers, to rebuild trust and heal some of the

Table 4.4 CB Insights' AI Fintech Market Components

1.	Automated credit scoring for direct lending applications
2.	Regulatory, compliance, and fraud detection for abnormal activities and behaviour, improved workflow
3.	Assistants or conversational AI in customer service and personal finance
4.	Quantitative and asset management algorithmic trading and investment strategies
5.	Insurance to quote and insure clients
6.	Market research/sentiment analysis
7.	Debt collection – personalised and automated communication
8.	Business finance and expense reporting
9.	General purpose/predictive analytics for general purpose semantics in customer care, and broad business intelligence and reporting

pain caused by the 2008 financial crisis. These are big shoes to fill. Everyone involved in bringing fintech about should never forget what's actually expected of fintech. If fintech is to achieve anything of substance, it has to start redefining trust with customers, so deploying ethical AI is vital. And so, in 2019 when questionable business practices by some of the darlings of fintech surfaced in the media, the entire fintech ecosystem suffered, stalled, and succumbed a little.

We simply cannot afford to make the same mistakes in fintech–to abuse systems, regulations, and use technology to cut corners. The investors in fintech need to have a voice in correcting questionable business practices by the fintech start-up's founders or the management, because without their money, fintech wouldn't be able to advance and grow. Over and above making money, the fintech investors are the ultimate guardians of fostering a healthy fintech ecosystem that will deliver value and trust and not inflict more pain on customers, erode trust, and undermine financial stability. Exactly as Environmental, Social, and Governance (ESG) is on the top of the agenda of every board, driving a close inspection of how investments are made to foster ESG criteria, investments in fintech and ethical AI technologies will be closely scrutinised. Whatever you pour money into grows. Let's pour money into something worthwhile with a vision well beyond merely making money.

Fintech growth will make financial services an even more competitive space. Emerging technologies will help cater to customer needs, which have never been catered to before, inefficiencies will be reduced in an attempt to maintain a competitive position, and new revenue and operating models will emerge. Such profound changes will also create new risks and have an impact on how regulators will deliver their role, to better manage a new order in financial stability.

In the UK the fintech sector is highly engaged with technology and is considered one of the most dynamic in the world. In a few years, we have seen incumbents and start-ups merging through strategic partnerships, investments, or simply buyouts. For instance, in September 2018 Barclays announced that they were using Simudyne, and in March 2019 they invested $6 million in Simudyne, an AI company specialising in computational simulation. In early 2018, Bank of England found Simudyne solutions of use, too. Scenarios simulation with agent-based modelling is a valuable tool to help with improved decision making. In retail banking, NatWest announced in August 2019 that they will begin trialing voice-only banking. This is a first by a UK bank. NatWest's customers will be able to bank (make payments, transfer money, pay bills, etc.) by talking to Google Home smart speakers.[153]With a number of well-run fintech incubators and regulator sandboxes, the fintech sector in 2017 contributed 76,500 jobs to the UK economy and received US$1.8 billion of venture capital. In 2018, this trend continued. In addition, there was a UK-based payment processing fintech acquired for US$12.9 billion, which counted for the largest fintech merger and acquisition deal in the world. In 2019, Monzo Bank announced series E funding of $200 million, which values the challenger bank at £2 billion up from £1 billion in October 2018. In 2018, the UK-based fintech ecosystem received the largest round of venture capital fintech funding in Europe.

Against this backdrop, the newly formed and quickly evolving AI fintech ecosystem is an exponential magnification of everything that fintech can deliver. In simple terms, the AI ecosystem is formed by (1) the incumbent financial services (the users of AI technologies), (2) the AI start-ups (the vendors of AI technologies) and (3) fintech firms as they are moving into using AI technologies to improve their offering. With fintech start-ups mainly using SaaS, the buyers are rather well versed in what they are buying or investing in. With AI fintech, the situation changes fundamentally. The issues arise when the buyer does not understand what they are buying and when the sellers mislabel their products as AI. Let's take a closer look at the current buyers/vendors interplay.

The decision to deploy AI in the enterprise is driven either by (1) a clear business direction to use AI as a tool for business growth and embed it into business strategy or (2) just an experiment of mythical shape. The users either buy AI or they build it. Now, let's zoom in on the buy scenario. The motivation for purchasing AI seems to be driven by four main forces:

1. "Silo" Outlook: AI tools are bought to solve a problem in isolation.
2. "Toes in the Water": AI tools are bought to "experiment with AI" or simply deal with "the low hanging fruit," and you can hear teams describing their AI aspirations.
3. "We Need to Do Something about AI": AI tools are bought as the result of a confused leadership, which can only give a confusing direction. For this purpose, usually anything that is remotely labelled as AI does it, and the board will be duly informed that the business is doing something about AI.
4. " Joined Up" Board Decision to Reinvent Their Business: This tends to carry weight and business strategy considerations; they go to market and buy what they need and with the long-term vision of tapping into next generation AI technologies and transforming their current business model.

These forces change depending on the type of leadership team: (1) visionaries, (2) experimenters, or (3) laggards.

The AI vendors fall into two main categories:

- vendors of real AI products
- vendors of pseudo-AI products

The users' or buyers' lack of education and knowledge sometimes set them down the path of buying pseudo-AI products. There are many people who have raised concerns about the practice of vendors/software companies promoting themselves as AI companies. It misleads the market, undermines trust, sets a wrong precedent, and is frequent enough to raise legitimate concerns.

ROBOTIC PROCESS AUTOMATION (RPA) IS DEAD

RPA and its recent reincarnations–Intelligent Process Automation (IPA) and smart RPA–are intricate names for what I call *linear automation.* This is a *task automation* tool of manual routine *within* the confines of *independent data and task silos.* It does not have the capability to work with unstructured data. Usually the best candidates for linear automation are repetitive–manual and menial tasks performed by humans.

RPA software does not learn either from the data it uses or from the feedback it is given by humans. It does not provide insights or adapt. RPA increases productivity by performing repetitive tasks faster at a reduced cost, but it is limited to the silo structure into which it is deployed. In contrast, AI provides what I call *exponential automation*. This is a *task automation* tool that uses *adaptive intelligence across data (including unstructured data) and task silos.* AI learns from its work, adapts in order to optimise its task, and provides valuable insights in real time, drawing from access to data across separate silos. If set up correctly, it is not limited to a silo structure.

Table 4.5 The Benefits of RPA

There are some notable benefits of RPA, which are summarised, quoting The Lab Consulting analysis:

1. Increased accuracy, reduced errors
2. Boosted productivity
3. No changes to existing IT infrastructure or applications
4. Quick installation and configuration
5. Reliability
6. Reduced processing cost
7. Reduced operational risk
8. Improved system data quality
9. Improved customer journeys and experience
10. Improved employee morale
11. Improved regulatory compliance adherence
12. Low technical skill barrier to entry

RPA has had a huge success. In a soup of acronyms–IPA, APS, IA, AIA, DI[154] and many more that are the product of hard work in the marketing departments of large consultancies–RPA has been the constant on which many organisations started their digital journey. These organisations were sold RPA as an intrinsic element of cognitive computing, which is another marketing gimmick term to create confusion and unduly shine to justify exorbitant prices for linear automation products.

It's no surprise that I found it difficult to find a natural home for RPA in this book. RPA shouldn't be included under the AI, cognitive, or intelligent banner. RPA has nothing to do with any of these, not even under its new name: Intelligent Process Automation (IPA). However, I decided to discuss RPA to dispel the misconceptions about a product that our industry has been adopting as if it the answer to their business transformation agenda. It is not. Herd mentality was a core driver of this adoption–if my peers are buying RPA it means that it must be the right thing. RPA companies' sales and marketing teams' creativity was another driver.

The year 2019 shone a new light on RPA. Companies like Symphony, an RPA provider and adviser, started rebranding themselves as "AI and RPA" in 2017 and even built an "AI lab" to support this rebranding. Shortly after the rebranding, they sold their business. In 2019, Blue Prism partnered with Cognitive Scale and Kore.ai as a strategic move, recognising that they need to increase the scope and capabilities of their RPA product. The RPA providers have started to quietly admit that they cannot upgrade their products to match what true AI technologies deliver. So, against this backdrop, the RPA providers need to keep the music on and so they have started splashing on PR. In 2019 BluePrism's annual event took place at a stadium–they needed that much space to host their developers and community with a lot of entertainment and partying. They had that much cash to throw around. I do not want to be a killjoy, but all of this is ephemeral. When companies begin splashy PR campaigns, it's a sign that they do not have a lot of substance, so they have to sell an image instead. I am not critical of what RPA actually is–from a technical standpoint it is a mere linear automation–but I *am* critical of the marketing teams and boards of RPA companies that knew exactly what they were selling and yet didn't hold back from up-selling.

RPA space is a lucrative one. Naturally, RPA experts defend their space at any cost. The best way to explain this would be through what I experienced first-hand when I approached a known London-based researcher in RPA and invited her to review an early draft of this book. She then rudely retorted her dismay at the lack of foundation for my strong views. I can only assume that this section struck a nerve. She then followed by a prompt blocking of my social media accounts. This reaction is both startling and fascinating in equal measures - only time will tell if this was a reaction of professional insecurity or a sense of her self-perceived academic superiority, although I would lean towards the former.

On 14th May 2019, I was invited to keynote an audience predominantly formed by the RPA community. During the fireside chat, I was asked if I concur with the analysts' recent stance that "RPA is dead." Yes. For the past three years I have been actively discouraging clients from buying it and investing in training and infrastructure to support RPA deployment. When your competition is investing in *fast cars*, why buy a *tricycle*? There is also another point: The infrastructure needed for a tricycle is not suitable for a fast car. In

2019, RPA/IPA companies commanded eye-watering valuations, and venture capitalists are patting themselves on the back about how well their investments in RPA companies are doing and how RPA companies cash in revenues beyond imagination. In 2019, UiPath is a $7 billion valuation company which has shot into the valuation stratosphere in the space of 24 months. Is this for real and for long term? These valuations may be sustainable if these companies do a U-turn and shift into AI. Daniel Dines, UiPath's CEO and founder, would be perfectly able to achieve such a U-turn, likewise with the CEOs of other RPA firms.

The RPA products that these companies sell are quickly becoming legacy for banks–a clunky software that they'll need to get rid of in a few years' time. Was that a long-term investment? No. HFS Research is a Cambridge, UK-based research house with offices in the US and India. They produce quality reports that would appear impartial. They seem like a valuable source of information on RPA and process optimisation vendors.

The hype and optimism around RPA has hit a halt, even according to RPA companies' social media accounts. As RPA companies explain there is a "gap between expectations and reality." Is there? Who created these expectations in the first place? Surely not the clients who bought RPA products.

ACCELERATING CHANGE

Financial companies access technology firms through partnerships, buying AI systems, investing in AI companies, or buying them out. This blueprint will remain the same for the AI fintech ecosystem. More AI companies will be created and will enter the already-established network of those who are accelerating the change, seeking to disrupt themselves from within or search for investing in tools that they can use and also sell to their competitors, as is the case of Blackrock's Aladdin.

Some financial companies have created their own accelerators, thus reaching out to the entire market. Wells Fargo has a competitive Start-up Accelerator program that has received more than 1,100 applications from over 50 countries since its inception in 2014 and its own Innovation Lab. Standard Chartered opened in April 2019 and launched their Africa eXellerator innovation hub in Kenya. Citi Ventures has an international global network of tech companies that participate in its six Citi Global Innovation Labs. These programs aim to build a portfolio of start-up investments, with a wide reach from eCommerce to cybersecurity. Axis Bank, India's third-largest private sector bank, launched an innovation lab called "Thought Factory" last year to accelerate the development of innovative AI technology solutions for the banking sector. In 2018 a regulatory sandbox, Fintech Hub, was set up at the Bank of England. Its focus is "the stability of the financial system, the safety and soundness of financial firms, and the bank's ability to use technology to supervise firms." Shearman & Sterling LLP have produced a very useful interactive map of all fintech regulatory sandboxes from different jurisdictions across the globe. Their work is highly relevant, because in time, these centres will become AI fintech hubs, leading work in AI developments in financial services.[155]

Through Citi Ventures, CitiBank has made a strategic investment in Feedzai (a leading fraud-prevention AI company) and Clarity Money (a personal finance company). Société Générale set up their venture fund in 2019, a cross-business innovation fund of €150 million that invests in internal and external start-ups. This fund has a clear mandate to support business development through their investments, with a particular focus on payment and e-commerce services, open banking, and small- and medium-sized enterprise services. Société Générale Ventures aims through their investments to optimise and create synergies across the group. Fidelity investments was one of the three leading investors in 2018 in a Series C+ round $620 million investment in the Chinese biometrics company SenseTime, founded in 2014. They are the leader in next-generation AI in intelligent vision, leading in computer vision applications used in surveillance, such as face, text, human and object recognitions, vehicle identification, and image-processing technology solutions, but also used in mobile internet, online entertainment, automobile, finance, retail, education, and real-estate industries in Hong Kong and China.[156] According to the database company Crunchbase, this is the most well-funded AI start-up to date, with a reported $1.6 billion in funding.

An example of protecting the current market position is Lloyds Banking Group investing in Thought Machine. Lloyds aims to address their own legacy challenges, and they announced a strategic partnership with the London-based Thought Machine that developed Vault, a cloud native modern alternative to the legacy platforms in banking,[157] which is a mix of smart contracts written in code, digitally signed by the customer and bank, and which also uses data-streaming architecture managed by machine-learning algorithms trained for this data warehouse. Lloyds has made an £11 million investment in Thought Machine, representing a 10 percent stake, as part of its "Series A" £18 million investment round, and Lloyds is open to continue finding suitable technology companies to help them deliver their three-year strategic plan.[158] In insurance, according to a Crunchbase report as of 11 April 2019, the top five AI insurance companies in 2018 were Lemonade, Artificial Labs, Flyreel,

Weather Check, and AI Insurance. Allianz was one of the investors in Lemonade's $300 million series D investment round lead by SoftBank.

Competing with financial services investment arms, in the small pool of good AI companies, there are the established VC and technology companies. Let's take a look at the most active investors by the number of investments in the top 100 AI start-ups in 2019, as identified by CB Insights, which also identified over 680 unique investors who funded the top 100 AI start-ups. Out of the 100 AI start-ups, 23 were listed outside of the US.

AI COMPANIES IN FINANCIAL SERVICES

According to MMC Ventures 2019 AI report, the UK remains the powerhouse of European AI entrepreneurship with 479 start-ups, followed by France (217), Germany (196), and Spain (166). In 2019, talent is scarce in this field. AI requires advanced knowledge in statistics, mathematics, and programming. In time, however, it is expected that there will be dedicated AI systems that will make AI accessible to those who are less specialised. However, the AI talent gap remains to be addressed. There are AI centres of training either online or in person. AI centres of training online are vastly accessible via Coursera or other online AI courses. In Barcelona, Jan Carbonell founded TheAcademy.ai to address the lack of AI skills in Spain. It has been reported that in 2019 the technology sector and financial services took close to 70 percent of the AI talent.

In financial services, this trend will continue to build up in the fintech ecosystem. In time, the vast majority of companies in financial services will likely be using AI technologies. Fintech companies are just a precursor to that stage.

As a snapshot of the most popular fintech companies, we should take a look at the following list (includes AI-first companies), which is a way to demonstrate that fintech companies are becoming AI-first companies:

Table 4.6 Most Popular Fintech Companies Use AI

Fintech Banking with AI virtual assistants	Ally Assist by Ally; Eno by Capital One Erica (Bank of America); IPSoft; Soul Machines Personetics; Clinc
Fintech Banking: fraud detection and security	Feedzai; InAuth; Simility; Swift
Fintech banking: neo-banks and challenger banks	Aspiration; Atom; Chime; Empower; Fidor; Monzo; Revolut; Simple; Starling; Tandem; Varo; N26

Fintech banking: mobile payments	Ayden; Zelle; Square; Cash by Square; PayPal; Venmo by PayPal; Apple Pay by Apple; Google Pay by Google; Samsung Pay by Samsung; Finn by JP Morgan Chase; Greenhouse by Wells Fargo; Marcus by Goldman Sachs; Samsung Pay; Vocalink by Mastercard
Fintech finance: lending	Affirm; Avant; Bond Street; Braviant; Credible; CrediFi; Earnest; FundBox; LendingTree; OnDeck; OppLoans; PeerIQ; Prosper; Salt Lending; Tala; Quicken Loans; Rocket Mortgage by Quicken Loans; LoanBuilder by PayPal; Gradifi by First Republic; Marcus by Goldman Sachs; Credible Labs
Fintech finance: crowd-funding + freelance market	CircleUp; Experiment; Fiverr; FloShip; Fundable; FundRazr; GoFundMe; IndieGoGo; KickStarter; MedStartr; Patreon; PledgeMusic; Pozible; Thrinacia; Ulule; Zopa
Fin-tech services: business solutions	Cambridge; Cambr; GeoTab; Global Payments; BrainTri; Finantix; Jellyvision; MuleSoft; Numerated; Plaid; Rein
Insure-tech services: insurance for consumers + business	Beam; BoughtByMany; BuzzVault; Corvus; Cuvva; Friendsurance; HavenLife; Hippo; Lemonade; Layr; Neos; Pluto; Sure; Trov; Zensurance; o2 Drive by o2

INVESTMENTS IN COMPUTING PROCESSING POWER

"AI does not stand still" Nvidia's website announces to its visitors. The enterprises that will be the long-term winners are those that don't stay still: (1) invest in the next generation computing (i.e. high-performance computing companies), (2) design their enterprise transformation as a full AI-stack approach rather than piecemeal digital transformation projects, and (3) have a clear strategy to invest in their processing power and possibly even dare to consider building independence from cloud providers. This framework applies to AI start-ups and incumbent financial services alike. An example discussed earlier in this chapter, *SenseTime*, is the best-funded AI company in the world as of now[159] for many reasons. One reason is that they have designed their business model as a full-stack AI.

SenseTime has their own independent, large, high-performance computing network that supports its own research and development and all of its AI applications. According to a report by Gregory C. Allen of the Center for a New American Security, SenseTime's computing network includes an impressive "54,000,000 Graphical Processing Unit (GPU) cores across 15,000 GPUs within 12 GPU clusters."[160]

Processing power is the engine that enables AI to work. Open AI, a US non-profit organisation, notes the demand for "compute" (computer scientists' jargon equivalent for processing power), for the AI projects have been doubling every 3.5 months since 2012. In other words, without adequate compute there will be a bottleneck in progressing with AI.

In Chapter 3, I discuss the taxonomy of different chip technologies. There are currently three categories mainly in use (1) Central Processing Units (CPU); (2) Graphics Processing Units (GPU); and (3) AI chips. New Street, a research firm, estimates that the market for AI chips could reach $30 billion by 2022. Intel is expected to earn $22 billion of revenue per year from selling processors for server computers. New Street expects that "ever more demanding AI workloads needing special treatment, fast-evolving algorithms, and tech giants designing their custom chips all may lead to a world in which lots of processor architectures thrive."[161]

Depending on your AI project and final objective, you might be using GPUs to train a large neural network in deep learning (for taxonomy of machine learning, see Chapter 2), but it is quite possible that a CPU is what's needed once the model is put in production (in use). GPUs are also used in cryptomining, which is particularly demanding of processing power.

In understanding the current state of AI, executives also need to understand where the semiconductors[162] market is going, as these signals should influence their investments in AI for business growth. This section offers a cursory look at how this market is trending and its independent, biggest,[163] and emerging players. There is a trend of fragmentation in the silicon market. The need for compute is higher for every AI project. AI-first companies like Apple, Alibaba, Tesla, and Baidu are building their own AI chips to break away from dependencies on the traditional suppliers. In addition, government initiatives like those in China set new dynamics that impact traditional players. The Chinese government announced that they will spend tens of billions to create a national semiconductor industry in order to also break away from Western imports. As cloud computing providers like Microsoft and Google morph into AI-as-a-service providers, their need for processing power is ever growing and they are developing their own computing infrastructure. Algorithmia is the only major vendor that supports serverless execution (FaaS) on GPUs, and it is useful for scaling certain AI projects.

Table 4.7 Semiconductor Companies

ESTABLISHED PLAYERS	
Nvidia	Cuda, Nvidia's platform, a GPU-based computing, is often referred to as the only suitable for machine-learning applications. Nvidia invented the GPU technology, which was originally used in gaming and is excellent for fast parallel processing of simple operations. GPU is now also used in cryptocurrency mining, training artificial intelligence systems and autonomous cars. Nvidia graphics processing units (GPU) have been widely adopted for work in deep-learning projects. A measure of its success was that in 2017 Nvidia stock increased 224 percent.
Intel	Intel is best known for the central processing units that power Windows PCs and Macs, but its other chips can be found in iPhone modems, data centres, drones, and self-driving cars. Intel manufactures and designs chips. A chain of acquisitions of Altera ($17 billion acquisition), Nirvana ($400 million acquisition), and Mobileye have expanded its capabilities to power autonomous vehicles' processors, servers processors, networking, and AI.
Qualcomm	Qualcomm's heritage is in inventing 3G and 4G cellular networking technology and they have invested in building and successfully delivering mobile processors used in Samsung and Apple, until Apple stopped using them. In 2017 Broadcom announced that they plan bidding $100 billion for Qualcomm.
Lattice Semiconductors	They attempted to be acquired by a China state-based company for $1.3 billion but President Trump blocked the deal over national security concerns.
Arm Holdings	A UK-based chip designer at the heart of most phones was acquired by SoftBank of Japan for GBP 34 billion
NEWCOMERS/ INDEPENDENT COMPANIES	
Imagination Technologies	Acquired by a China-backed private equity firm for GBP 550 million
Graphcore	A UK-based semiconductor company that develops accelerators for AI and machine learning have raised series B for approx $300 million

Hadean	A start-up in London that works with a number of customers in finance, genomics, and gaming. They are still trialing in beta and have re-engineered the entire computing stack to put supercomputer levels of processing power at the disposal of anyone with a laptop and internet access. This could be transformative.
Cerebras	A US competitor to Graphcore
Cambricon	A start-up based in Shanghai, recently unveiled a chip that is similar to Graphcore's and Cerebras's. Their development is keenly supported by the Chinese government.

AN AI APPROACH TO INVESTING

In 2014, VITAL was presented as the "first board director" with a board seat and a vote in the investment management committee of Deep Knowledge Ventures, a Hong Kong investment company specialising in biotechnology investments. VITAL stands for Validating Investment Tools for Advancing Life Sciences.

Deep Knowledge Ventures (DKV) claims that they avoided bankruptcy[164] thanks to VITAL, the first artificial intelligence biotech investment analysis tool. VITAL enabled DVK to identify more than 50 parameters that were critical for assessing risk factors. VITAL also showed the probability of success in certain subsections, correctly predicting signals and paradigm shifts that would benefit DVK. VITAL 2.0 was launched in 2017, and the company does not make any investment decisions without corroborating them with VITAL's findings.

Hone Capital is the Silicon Valley-based office of one of the largest venture capital and private equity firms in China. For their US operations they have developed a data-driven approach to analysing potential seed deals, with promising early results. They use machine learning to mine large quantities of data from their strategic partner AngelList, an angel-investing platform. They started using more than 30,000 deals from the last decade, using data from various sources like CrunchBase, Mattermark, and PitchBook. Their algorithms are able to identify the probability of success of a start-up. They identified cases where the algorithm was able to compute data that the investment team wasn't able to, and it would miss out on deals.[165]

ACADEMIC AND INDUSTRY AI RESEARCH

The brain-drain from academia to industry will have likely implications on slowing research and teaching, but it will move value into catalysing AI's immediate impact. The leading global banks have understood the importance of research, and they are now focusing on building their own research labs. For instance, the JPMorgan AI research team is committed to building a reputation as Wall Street's leading technology bank. In mid-2018, they announced that Professor Manuela Veloso, formerly Head of the Machine Learning Department at Carnegie Mellon University (US), joined them in a full-time role[166] to run their corporate and investment banking AI research. "Providing easy-to-use technology in order to deliver a great client experience will continue to be a major differentiator," according to Daniel Pinto, co-COO of the investment bank.[167] JP Morgan's strategy is also to partner closely with universities and research institutions in this space[168] in search of furthering opportunities and designing solutions that can potentially transform financial services.

Royal Bank of Canada (RBC) is Canada's largest bank, and one of the largest banks in the world, based on market capitalisation, with 81,000 full- and part-time employees and 16 million clients across different categories. Borealis AI, the RBC Institute for Research in Artificial Intelligence, was established in 2016 and aims to build state-of-the-art machine learning focusing on ethical AI. Their AI experts are world-class and they are led by Professor Matthew E. Taylor, formerly with Washington State University (US), who relocated to Edmonton, a leading centre in reinforcement learning, a machine-learning approach that is revolutionising learning. The RBC website announces that "Prof. Taylor adds to the critical mass of experts currently relocating to the area." In 2018, Borealis AI opened a new research centre in Vancouver, which focuses on computer vision. Professor Greg Mori, director of computing science at Simon Fraser University and an internationally recognised expert in computer vision, heads research efforts at the lab as research director. Professor Mori continues to teach at his university, but obviously on a reduced schedule.

According to *The Economist,* between 2006 and 2014, there was an increase from just under 2 percent to almost 40 percent in AI research publications, including an increase in authors with corporate affiliations. In 2017 and 2018, several leading AI researchers moved to technology firms for significant salary increases. With such substantial resettling in the industry, evidently there's less teaching staff available to teach the next generation of academics, which will only exacerbate the lack of AI talent.

Fifteen or 20 years ago, academia was able to retain the brightest minds, especially those who would be attracted by an investment bank. Today, technology companies like Facebook, Apple, Google, Amazon, Microsoft, Uber, and others offer a highly stimulating research environment and ample resources, which are always an attraction for academics, as such an environment frees them from filling out the time-consuming grant applications to continue their research. This further limits the pool of talent in AI research.

Uber hired 150 research scientists from the US National Robotics Engineering Center based at Carnegie Mellon University in Pittsburg. In the UK, the University of Cambridge hired Zoubin Ghahramani, one of the most respected AI researchers in the country, to become chief scientist and head of machine learning.

Google hired Geoffrey Hinton, a deep-learning pioneer at the University of Toronto, in 2013, while he was still teaching. In 2016, Google DeepMind developed a research partnership with Oxford University where they fund more than 250 research projects and a dozen PhD scholarships. In 2018, DeepMind donated an undisclosed amount to the University of Cambridge to appoint a "DeepMind Chair of Machine Learning," possibly as a reaction to being criticised for recruiting academics as full-time employees. A revolving door between academia and industry is healthy and welcome–within reason. Yoshua Bengio, who advises IBM and Microsoft, thinks that the best research labs will attract the best talent. This becomes the catalyst for wealth and power in the hands of the few and this is "dangerous to our society." A consequence of AI technology is a winner-takes-all approach.

While North America is leading in the number of AI companies, according to *Times Higher Education,* in the period between 2011 and 2015, China published more than 41,000 papers on AI. That's almost twice as much as the US number. In 2019, the Allen Institute for Artificial Intelligence created a tool called Semantic Scholar, which uses AI to search and analyse scientific papers published online. The study suggests that China will overtake the US in the most-cited (50 percent) research papers in 2019, the top 10 percent of research papers in 2020, and the top 1 percent by 2025.

The list of exceptional contributions to the field of AI research from academia is long. In addition to the University of Oxford–notably Oxford-Man Institute of Quantitative Finance–and the University of Cambridge, it includes, among many other distinguished academic institutions, which can always be regarded as potential partners for financial institutions:

- MIT, with the Department of Brain and Cognitive Sciences; MIT Schwarzman College of Computing and MIT Media Lab.

- Carnegie Mellon University, with their world-distinguished Mellon College of Science.
- Stanford University, with The Stanford AI Lab, which provides individual courses as well as brings together students from different faculties and disciplines.
- University of California, Berkeley, which has the renowned Berkeley Artificial Intelligence Research Lab.
- Nanyang Technological University in Singapore, which is pushing the frontiers of AI with their research labs at the smart campus.
- Harvard University, with the John A. Paulson School of Engineering & Applied Sciences (SEAS) and the Institute for Applied Computational Science, which provides a varied selection of programs specialising in AI.
- University of Edinburgh, which has a long history in researching AI even before it was a white-hot topic; their five-year, full-time degree provides deep studies in computer science AI, cognitive science with linguistics, neuroscience, psychology, and biology.
- Queen's University Belfast, a hidden gem of talent and quality of teaching in AI.
- University of Lugano, which made a historic leap in deep learning with the work of Jürgen Schmidhuber, who invented the long short term memory (LSTM) algorithm with applications in predictive texting and others in smart phones.
- University Paris Saclay, UPMC, CentraleSupélec, the French Institute of research and higher education in engineering and science.
- The Jožef Stefan Institute in Slovenia (EU), a too-little-talked-about AI research centre that has delivered outstanding AI projects more recently with Japanese researchers to design evolutionary algorithms that combine with visualization tools and real-world applications.
- National Kaohsiung First University of Science and Technology in Taiwan, due to the work of Tung-Kuan Liu, Ph.D.
- Ritsumeikan University Japan, for the much-cited research by Professor Joo-Ho Lee, Ph.D.
- Zhejiang University of Technology China, famous for the work of Qiang Chen, Ph.D. in AI and industrial automation.
- Politehnica University of Timisoara, Romania, for the work of Professor Radu Emil Precup Ph.D. in the development and analysis of new control structures and algorithms.

When talking to various scholars, I have been able to see a common thread they all subscribe to: the need for an open field in AI research, more connectivity, and visibility to share one another's work. This would require substantial and state-level coordination. While it would help to have such coordination, I think an enterprise-level and industry-level coordination would be beneficial.

GOVERNMENT ROLE IN AI DEVELOPMENT

Important academic research has gone into analysing how the government's role, both in direction and funding, has shaped how AI technologies have grown. Professor Mariana Mazzucato, the Chair in the Economics of Innovation and Public Value at University College London (UCL), has analysed the public policy in shaping both the rate of growth and its direction. Her book *The Value of Everything* is an informative read.

Professor Mazzucato's point is that businesses tend to invest in technological development only when they see profits, whereas governments can take a more strategic view, not driven by immediate profit-seeking. One of the original engines of Silicon Valley's creativity, success, and profit, she argues, was the Defense Advanced Research Projects Agency (DARPA), founded by President Dwight Eisenhower in 1958, following the alarm caused by the Soviet Union's launch of the Sputnik rocket. DARPA, run by the US Department of Defense, has since spent billions of dollars on cutting-edge research and was instrumental in developing the internet,[169] alongside other technologies that were accessible only to military special forces.

The AI winters that I have discussed were driven mainly by the US and UK governments pulling the funding of what was once considered pure academic curiosity. The Western world used to lead in shaping the developments in AI. Since 2017, China's AI strategy is one of the most visionary and committed to investing in all preconditions of AI because "Beijing Wants A.I. to Be Made in China by 2030" in the words of *The New York Times*.[170]

EXAMPLES OF QUESTIONS FOR BOARDS TO ASK

	TOPIC	POSSIBLE QUESTIONS
1	Investing in AI	1. When investing in an AI company, which component is material to your business: their (1) hardware, (2) software, (3) data, or (4) data science talent? 2. What problem does this use of AI solve? How do you measure outcomes? 3. How does the AI provider capture its share of value? What is their market reach? 4. How is this company valuable to our long-term growth? 5. What is the long-term social impact of investing in these AI companies? 6. Have we carried out a thorough due diligence? Have we covered all points discussed earlier in the chapter? 7. Are we investing in companies that are likely to infringe on human rights like privacy? 8. Are we investing in infrastructure companies that will enable us to take an improved look at our data strategy, improve our AI infrastructure?
2	Research and Development	1.Have we any plans to develop this capability either internally or externally through strategic partnership with academic centres? 2.Do we invest in fintech accelerators? 3. What type of engagement do we have with the fintech community? How about the AI research community?

KEY POINTS TO REMEMBER

- In order to succeed and grow, today's financial institutions must incorporate cutting-edge technology into their business strategies.
- If you choose to buy rather than build AI tools for your business, you must understand the difference between real AI products and pseudo-AI products that market their software as AI when, in fact, it does not use adaptive intelligence to optimise tasks.
- There are three AI evolution waves: past or *symbolic* AI, with manually input knowledge, which had strong reasoning but poor learning capabilities; present or *statistical learning,* which has strong learning capabilities but less strong reasoning; and future, described as *contextual adaptation,* which will be at the intersection of the two previous waves.
- The board and executive teams must have a framework to inform themselves about the particularities of due diligence in AI transactions, including areas such as intellectual property, data privacy, cybersecurity, insurance due diligence, AI governance, product liability, surveillance, foreign investment/national security, other regulatory requirements, and the company's executive management.
- RPA software does not learn either from the data it uses or from the feedback it is given from humans. It does not provide insights, does not adapt, and it does not learn from the data it uses.
- Many financial companies are starting their own innovation labs to accelerate change.
- The enterprises that will be the long-term winners are those that: (1) invest in the next generation computing (i.e. high-performance computing companies), (2) design their enterprise transformation as a full AI-stack approach rather than piecemeal digital transformation projects, and (3) have a clear strategy to invest in their processing power and possibly even dare to consider building independence from cloud providers.
- In understanding the current state of AI, executives also need to understand where the semiconductors market is going, as these signals should influence their investments in AI for business growth.
- Some boards are using AI to help them assess risk when investing–some even have AI board directors.
- There is a growing lack of AI talent, as academics move to technology companies with ample resources.

CHAPTER FIVE

Corporate Governance and AI Adoption

This chapter addresses the intersection of various fiduciary duties that board directors have with deploying AI in their organisations. The governance of AI will remain at the forefront of the design and deployment of AI in financial services and will continue to closely scrutinise the use of AI. Therefore, it is imperative that boards and decision makers are aware of the current elements that form the narrative around the governance of AI–and the risks derived from adopting AI without ethical frameworks. This chapter aims to provide a snapshot of this narrative.

Five years ago no one really discussed AI ethics. Three years ago, we started implementing ethical evaluation frameworks for AI vendors and tools we would recommend to our clients. Allegedly, no other consultant was discussing ethics in technology with their clients. In 2019, big four consultants competed with each other to promote themselves as specialists in ethical AI. Twelve months ago, a board director asked me what an AI ethicist does.

Governance of AI is not compliance–it is corporate governance and good business. Boards will be decidedly responsible for how the right actions have been taken and their decision-making process needs to be articulated and competent. This chapter addresses the majority of nuances that executives ought to consider when selecting their AI strategy and the AI tools to be deployed, from important ethical guidelines to customers' privacy to the dangers of algorithmic bias.

The opening line of an important paper aptly titled "AI4People–An Ethical Framework for a Good AI Society: Opportunities, Risks, Principles, and Recommendations," led by a consortium of leading European universities, reads as follows: "AI is not another utility that needs to be regulated once it is mature. It is a powerful force, a new form of smart agency, which is already reshaping our lives, our interactions, and our environments." This draws emphasis to the potential of AI and the relevance it holds. There are a number of concerns surrounding "governance," which remains a relevant topic of debate among many academics. However ironic, this same level of debate is rarely transferred to those in positions of business decision making–for example, board members. The main concerns are addressed in this chapter through the lens of the financial services sector and its decision makers. As early as 2015, I proposed that the governance of AI rests on four core intertwined values:

1. AI needs to have a human-centric design.
2. AI design has to promote positive values.
3. AI needs to have a switch-off button.
4. AI needs to be explainable and the humans who build it and deploy it are accountable

If organisations are going to incorporate AI into their business strategy–as they should if you want to remain relevant and continue to grow–then it's critical that they adopt an ethical approach to AI, both for the good of society and your own economic safety and stability.

FIVE IMPORTANT FACETS OF AI GOVERNANCE

The debate surrounding the governance of AI is a construct of five general ideas, which are closely interlinked: opportunities, unintended consequences, philosophical implications, challenges, and academic research in AI governance.

Opportunities

The paper mentioned above in the chapter introduction introduces four main opportunities that AI adoption offers. I mention these as I share the view that "humans first"[171] should be the rule of thumb in all things AI. The quadripartite structure reflects the four fundamental points in the understanding of human dignity and progress:

1. Who we can become (autonomous self-realisation)
2. What we can do (human agency)
3. What we can achieve (individual and societal capabilities)
4. How we can interact with each other and the world (societal cohesion)

There is tension concerning the privacy around the collection and use of datasets. In financial services, this narrative is set to relieve some of this tension. The data is already collected and its visibility and access to it should be within easy reach. However, the financial services industry is already strictly regulated, giving deep meaning to the assumption that data should not be abused.

Unintended Consequences

Trying to effect change or taking a new road never comes without risk. AI use cases have already proven that it can be used to foster human nature or to augment human abilities. This possibility creates a range of opportunities and one of the greatest risks is actually underusing AI, meaning a host of lost opportunities. Furthermore, lack of technical knowledge at the decision-making level breeds a fear of being wrong and making the wrong decisions. A byproduct of fear is deliberate ignorance, which may lead to underuse of this technology. This will ultimately translate as a missed business opportunity. Another risk is the misuse of AI. Misuse is not always easy to identify when consequences are not considered from the outset.

AI can be used to deliver a lot of good. Examples include improving our offering to clients, working conditions for our employees, or unlocking pockets of business growth. However, it comes with a caveat if left ungoverned at any point. An example could be what AI has delivered for the advertising industry: It gives access to services at better prices, but when left unregulated, it is used to manipulate people's choices, not necessarily in their interest. Another example comes from the automotive industry and tracking employees' use of company vehicles.

When using AI to predict frequency of maintenance, the data revealed that drivers had different styles of operating the vehicles. This, in turn, had a substantial impact on vehicle-maintenance costs. The company was able to identify those drivers whose driving style caused increased engine wear and mechanical stress. The natural reaction was to embed these metrics in the performance evaluation of the drivers. What started as a predictive maintenance could have easily wound up as an extra metric to evaluate drivers. Some might argue that this would be a bonus insight, as it would help reduce a company's overhead through improved operating parameters, however, there is a strong argument that it was unfair to burden the drivers with extra evaluation systems.

Philosophical Implications

The impact of data on contemporary society is characterised by its sheer volume, and the velocity with which it transforms each sector in our industry. A report by The World Economic Forum published in 2017 indicates that "the world produces 2.5 quintillion bytes a day, and 90% of all data has been produced in just the last two years."[172]

This also shows that we are at a point in our history when we are forced to confront moral issues that have never been raised before. In the context of modern societies, I would echo opinions that moral philosophy and AI ethics share one common denominator: responsible AI development. Ethics done with introspection, as John C. Havens proposes, is what we need to do in order to understand correctly the most pressing ethical questions that humanity has ever been faced with. John means self-introspection for each of us individually and for the organisations that we lead. This is essential, as data scientists aim to be "building models for how the world should be."[173] AI technology has already caused a deep social impact and so it is the "first time in history where philosophy has become relevant."[174]

Challenges

On November 14, 2018 I attended a talk at the University of Oxford given by Allan Defoe, a leading professor on global politics and governance of AI at Oxford and Yale. He is a Director of the Governance of AI Program[175] at the Future of Humanity Institute where he works with Nick Bostrom to investigate the impacts of technology on society. I believe that our industry needs to be aware of the thinking and concerns identified by such scholars and experts. In his talk, Professor Defoe summarises near-term governance challenges in AI as:

1. Safety in critical systems such as finance, energy, transportation, robotics, and autonomous vehicles
2. Encoding values in algorithms for human resources, loan decisioning, policing, justice, and social networks
3. Impacts of AI on employment, equality, privacy, and democracy

He then carries on to identify *extreme* challenges related to near-term AI:

1. Mass labour displacement and he makes the distinction between (a) AI that substitutes human labour and (b) AI that complements human labour
2. Inequality of earnings of those who control datasets and AI tools compared to those who do not
3. AI oligopolies in strategic industry and trade, which holds implications such as the early entrant advantage held by Google and Facebook, and of taxations of these entities with the need to redesign laws to reflect how these entities operate and develop mercantile trade arrangements
4. Strategic nuclear stability with a focus on autonomous escalation and nuclear retaliation
5. Military advantage or the use of AI cyber attacks, intel, info operations
6. Accident/emergent/other risks from AI-dependent critical systems and transformative capabilities

7. Protect AI innovation from fearful backlash, protectionist, and clumsy regulations (He is right, and this particular point is relevant to our industry and how fear paralyses actions and fuels business mis-opportunities. I'll develop this point later in this chapter.)
8. Surveillance and control, which brings into focus the power distribution between those who are surveilled and those who do the surveillance (This draws into question the use of data in good faith and when it begins to encroach on civil liberties. The latter outcome shows that surveillance can easily become a tool for persuasion and repression. In a video interview for the Council of Europe in 2018, Professor Joanna Bryson said that one aspect of governance is that how data is being used can change rather quickly with an election or change of political regime. Professor Bryson reminds us that there is no longer "anonymity through obscurity" and that during the Holocaust more Jews were killed in the Netherlands than in Belgium because the Dutch were more organised with their data and could trace the Jewish population more accurately.[176])

Academic Research in AI Governance

In recognition of its importance, AI governance has quickly become the focus of many charitable donations to academic circles. In October 2018 alone, Google DeepMind committed a donation to four University of Oxford scholarships to support underrepresented groups in computer science, such as women or people from low progression to higher education families. This is a response to the need to build diversity in technology development teams, as an insurance policy against unintended consequences like gender or class bias. Michigan State University received $1 million from Manoj Saxena, the first General Manager of IBM Watson, to study digital ethics and ethical AI.

Stephen Schwarzman, the CEO and co-founder of The Blackstone Group, donated $350 million to Massachusetts Institute of Technology (MIT), for a program on responsible and ethical AI uses, which will start in 2019. MIT announced that the final endowment for this college will be $1 billion. The University of Oxford's Future of Humanity Institute has just received a £13.3 million gift (approximately $17.3 million) shortly followed by £150 million also from Stephen Schwarzman to investigate AI in the Stephen A. Schwarzman Centre for the Humanities.[177]

This sets the scene for financial services boards to consider their own contribution to academia's fundamentally necessary thinking for the blueprint of responsible AI. Wallenberg Foundation, in addition to other philanthropical initiatives, donated just under

$67 million to Umea University for a program led by the highly admired Professor and AI ethicist Virginia Dignum. The program will "examine methods and tools for ensuring that AI and autonomous systems are designed so they do not clash with human values and ethical principles."[178]

AI ETHICS PRINCIPLES

What is ethics and why is it so important to AI? Ethics is the study of what it means to do the right thing or "the set of behaviours that maintains a society," as Professor Joanna J. Bryson summarises it. It is a complex subject that has occupied philosophers and experts from other disciplines for hundreds of years. Ethical theory assumes that people are rational and make free choices, but neither of these conditions is always and absolutely true.[179] While there is much agreement about the ethical rules in general, when it comes to what ethics mean in the deployment of AI, the opinions are vastly different.[180] Immanuel Kant's opinion should perhaps remain a guiding light in our thinking ahead on how AI will impact the legal construct around how we, humans, interact with machines: *One must never treat people as merely a means to ends, but rather ends in themselves*.[181] But what is ethics in computer science? In the US, as early as 1997, the Software Engineering Code of Ethics together with the IEEE Computer Society produced a code of ethics for software engineer practitioners that also included policy makers, educators, and supervisors. The Code is seen as a living document and contains eight principles related to behaviour and decision making–and the obligations derived from this profession. It is founded on the "software engineers humanity, in special care owed to people affected by the work"[182] and invites them to root their judgement in the "public interest."

As a fundamental principle, good practice software engineering[183] should be promoted. The study of ethics for AI programmers and researchers might need to become mandatory–perhaps even a Hippocratic Oath for coders–as Microsoft suggests in their book *The Future Computed: Artificial Intelligence and Its Role in Society*.[184] It's all about what computers *should* do, and not about what they *can* do, thereby reiterating the commitment to a broad societal responsibility.

Manipulating Choice

Many of the ethical concerns that have been raised are related to manipulating people's choices. Using data and insights into people's behaviour have been identified as the core cause of democratic processes being swayed and manipulated. This is a malicious way of using data to influence people. While we do not want to see people's political views swayed by insidious techniques, there's one instance where *manipulation of people's choices* might not be such a bad idea.

The payments industry aims to provide frictionless payments, but I believe that a little bit of friction at the pay point can actually be beneficial to users. A friction in payment allows the scope to design algorithms to nudge people–or to even manipulate people's choices–so that they save more, to control their spending impulse (such as buying yet another pair of shoes) or to help them reassess their spending and savings patterns.

The credit card was brought to the market in the 1960s and it represented an innovation in payments. Half a century later, we can see what a successful innovation it has been. It helped people access capital to buy household or personal use items. The respective industries flourished at the expense of individuals piling up ever more debt. Some people do not know how to use debt properly. Reports show that in the UK in 2019, there are about 100,000 individuals in deep debt who consider suicide on a daily basis. I am a strong believer that building a stronger savings culture with AI will bring individuals long-term empowerment and control of their finances and ultimately their wealth.

Unconscious bias is built early on in life and that's where our episodic memory is formed, and it informs what we do with our finances. This is a conditioning that we are, in some instances, not even aware of. It takes time to identify it, and reframe that conditioning for people's benefit.

Early-stage progress is being made in this space by two companies: Pefin, a NYC-based company in financial planning, and Oxford Risk, a London-based company in risk evaluation and investments.

AI or Human?

In 2018, Google made the headlines by launching Google Duplex, a digital assistant that was convincingly human-like (verbal ticks included) that would call shops, hairdressers, and restaurants on your behalf. Shortly after this launch, Google also made the headlines for the lack of ethical considerations in the Duplex's design.

The concerns were identified as misleading for an AI to pretend to be a human, when the human on the other end of the line did not know that he or she was talking to a machine. Since then, Google has advised that they are designing a built-in disclosure feature, so the system is appropriately identified as a non-human.

The Ethical Toolkit

Global AI.org based in Austin (Texas) and The Institute for Ethical AI and Machine Learning in London have proposed in 2019 valuable ethical evaluations toolkits. These initiatives were pioneered early in 2018, by The Institute for the Future and Omidyar Network which announced the availability of The Ethical Operating Systems Toolkit[185] for tech entrepreneurs, founders, and CEOs. The toolkit helps technologists envision the potential risks and worst-case scenarios of how their technologies may be used in the future, so they can anticipate issues as well as design and implement ethical solutions from the outset. The toolkit is useful not only for software developers but also for decisions makers. It is already being piloted in a large number of tech companies, schools, and start-ups. It identifies:
Risk zones where unintended consequences are most likely to appear:

1. Trust, Disinformation, Propaganda
2. Addiction and the Dopamine/Attention Economy
3. Economic and Asset Inequalities
4. Machine Ethics and Algorithmic Biases
5. Surveillance State
6. Data Control and Monetisation
7. Implicit Trust
8. Hateful and Criminal Actors
9. Future-proofing strategies that help technologists:
10. Prioritise identified risks
11. Determine the biggest and hardest-to-address threats
12. Choose where to begin to develop strategies that will help mitigate risks

Privacy and New Revenue Models

The attention economy is mainly associated with the business model scaled by Facebook. It has proven a highly lucrative model. In 2019, Facebook announced major changes to address privacy concerns. One of the changes was to encrypt end-to-end messaging on all their messaging platforms by merging them into one. Naturally, this move implies a shift in their revenue-generating model to find a new source of revenue.[186]

New sources of revenue may be "the pivot from monetizing attention to monetizing the protocol," with the announcement of their digital coin. Facebook has some of the best minds in Blockchain and DLT, and they stand a good chance of making it work from a technical standpoint. We'll take a more in-depth look at privacy later in the chapter.

Ethical Guidelines

Professor Alan Winfield has aggregated a useful timeline of the main principles of robotics and AI, titled "An Updated Round Up of Ethical Principles of Robotics and AI"[187] that I recommend as an ongoing resource for consultation. The timeline starts with 1950 and Asimov's laws of robotics. Reading this timeline, you notice two main details:

1. The principles were originally conceived by a sci-fi writer, then by academia, then by the industry (Google, Intel), then by a government (EU). This is a reflection of how this technology permeated our collective consciousness from a mere sci-fi story to the reality of impacting our lives.
2. All of these principles place an individual's protection and safety at the crux. As our industry will be adopting AI at a larger scale, these principles will filter through and deliver what financial services has always meant to deliver: human-centred services to protect the individuals and their wealth.

Also in March 2019, the European Commission publicly launched Ethics Guidelines for Trustworthy AI, a global breakthrough in defining ethical AI by some first-class ethicists from academia, law, and the industry. This document highlights some core principles that I believe will filter through into enterprise AI. According to these guidelines, trustworthy AI should be

Lawful: respecting all applicable laws and regulations
Ethical: respecting ethical principles and values
Robust: from a technical perspective while taking into account its social environment

These guidelines are useful as they provide further details and introduce seven key requirements that trustworthy AI systems should have to make them human-centric–a core value in ethical AI design. For my readers' convenience I reproduce them below:[188]

1. *Human Agency and Oversight*: AI systems should empower human beings, allowing them to make informed decisions and fostering their fundamental rights. At the same time, proper oversight mechanisms need to be ensured, which can be achieved through human-in-the-loop, human-on-the-loop, and human-in-command approaches.

2. *Technical Robustness and Safety*: AI systems need to be resilient and secure. They need to be safe, ensuring a fall-back plan in case something goes wrong, as well as being accurate, reliable, and reproducible.

3. *Privacy and Data Governance:* Besides ensuring full respect for privacy and data protection, adequate data governance mechanisms must also be ensured, taking into account the quality and integrity of the data, and ensuring legitimised access to data.

4. *Transparency:* The data system and AI business models should be transparent. Traceability mechanisms can help achieve this. Moreover, AI systems and their decisions should be explained in a manner adapted to the stakeholder concerned. Humans need to be aware that they are interacting with an AI system and must be informed of the system's capabilities and limitations.

5. *Diversity, Non-discrimination and Fairness:* Unfair bias must be avoided, as it could have multiple negative implications, from the marginalisation of vulnerable groups to the exacerbation of prejudice and discrimination. Fostering diversity, AI systems should be accessible to all, regardless of any disability, and involve relevant stakeholders throughout their entire life cycle.

6. *Societal and Environmental Well-being:* AI systems should benefit all human beings, including future generations. It must hence be ensured that they are sustainable and environmentally friendly.

7. *Accountability:* Mechanisms should be put in place to ensure responsibility and accountability for AI systems and their outcomes. Auditability enables the assessment of algorithms, data, and design processes and plays a key role therein, especially in critical applications. Moreover, adequate and accessible redress should be ensured.

ACCOUNTABILITY

An increased focus on the use of data has led to championing "accountability" as an answer to big-data issues. This has superseded the previous solution of increased transparency. Experts have put forward various principles to enhance the accountability construct, and they concluded that there are three main dimensions to address:[189]

1. *Explainability* conveys that the algorithm should be explained.
2. *Declarability* means that in cases where metrics and weightings must be hidden to prevent system gaming, the existence of hidden metrics should be declared by the algorithm designer, as a percentage impact on the algorithm's output. For example, "hidden metrics represent 34 percent of the algorithm's points."
3. *Auditability* denotes that a third party under a Non-Disclosure Agreement (NDA) should be able to audit the algorithm.

There is a common discussion in data-science circles, concerning the trade-off between explainability and accuracy of an algorithms' outcomes. This means that some AI models do not permit explainability, unless we are prepared to compromise on their accuracy. However, let's not take this statement as a convenient explanation for why AI models do not need version control or cannot be explained. People often suggest that everything should be explained. Now, I'd like to pose two practical questions: (1) Do we really need to have explainability across *all* outcomes and models? Probably not; (2) Which business processes, deployed with AI, must be explainable? Careful statistical testing of a system might be another way to make sure that a system works in the way it is intended. Testing might be a better basis for trust in those cases where we do not really need explainability.

As a reminder, interpretability, transparency, and explainability are all important as an aggregate, as well as independently. They all carry different value propositions for the task at hand. We need to have clarity of purpose and common sense in order to decide when explainability is actually needed. Do we need to know how an autonomous credit decisioning tool operates before it makes errors at the expense of people's credit rating? Absolutely. Do we need to know how anti-fraud tools achieve close to 100 percent accuracy of outcome? Not as much, providing that they deliver on anti-fraud prevention and protection.

AI SAFETY

In 2016, The White House Office of Science and Technology issued a "Request for information on AI"[190] and stated a clear focus to address the alignment problem and "leverage AI as an emergent technology for public good and towards a more effective government." One point on their agenda related to AI safety and control. AI researchers are uncertain about how the field of AI will develop. According to the University of Oxford's Future of Humanity Institute,[191] there are many scenarios that might occur and associated risks for humanity brought by high-level machine intelligence and super intelligence. AI safety is an ongoing concern that needs to be addressed. We must not wait until human-level artificial intelligence is achieved–we should start now. For instance, let's look at the problem of accidents in machine-learning systems, defined as unintended and harmful behaviour that may emerge from the poor design of real-world AI systems. Problems may arise from:

1. wrong objective function, such as avoiding side effects;
2. technology that is expensive to evaluate frequently; or
3. undesirable model behaviour during the training.[192]

A breakthrough approach to evaluating the trustworthiness of AI systems was announced by Cognitive Scale, the leader in practical, scalable, and trusted AI software. Their new product, Cortex Certifi, uses AI to detect and score vulnerabilities in almost all black box models without requiring access to model internals. Cortex Certifi can answer "Why this outcome? Has the model been biased? Has anyone been biased against?"[193]One proposal is to address AI safety by embedding it in the design and to think of it as a business case for good corporate governance, much like corporate social responsibility. As yet, we do not know what this will cost. It could be 10 percent of the development cost, or maybe 50 percent to start with, following which the cost will taper off. It is like building a bridge or manufacturing a new drug. Part of the research and engineering will be distributed over the life of the product/drug/AI/software.

There are a few places in this book where I mention the importance of weaving computer science, humanities, and responsible AI design in one degree. Such an example is Carnegie Mellon University's School of Computer Science, which in 2018 launched a Bachelor of Science in Artificial Intelligence (BSAI) that covers math, statistics, computer science, AI, science and engineering, humanities, and art.[194]

BIAS: DATA DOES NOT FORGET

There is substantial research and public discussion on "algorithmic bias." From the algorithms of leading AI firms unable to recognise skin colours other than white[195] to incorrect credit decisioning or court order rulings against people living at certain postal codes, bias discrimination is well documented. However, more importantly we need to answer the following question: Why do these errors happen and how can we correct them? The core issue is that algorithms are trained on past data. There is a high likelihood that this data reflects biases through historical prejudices, and so as we automate decisions using this data, we automate bias. As mathematician Cathy O'Neil points out, "Algorithms are opinions embedded in code." She summarises how the outcome of algorithmic decision making is not a scientific formula, rather the coders' own opinions become embedded in what the optimal outcome or success should look like.

In very simple terms, algorithmic biases could:

1. be hard coded by the developer
2. come from biased data
3. derive from a biased choice of features analysed and how we frame the problem to solve
4. arise from *overfitting*, otherwise known as incorrect correlations, a bias that is less talked about (overfitting occurs when a predictive model is so "specific" that it may not yield accurate predictions for new observations. To counter this, cross-validation techniques are applied.[196])

The concept of "biased data" is often too broad to be useful. In January 2019 two MIT professors, Harini Suresh and John V. Guttag, in their paper titled "A Framework for Understanding Unintended Consequences of Machine Learning"[197] introduced a clear framework to understand the five types of data biases and a useful explanation of their remedies:

1. *Historical Bias:* historic data reflects past patterns (e.g. when you search images for CEO, your search returns a majority of male images)
2. *Representation Bias:* data reflects one specific category missing the large majority (e.g. when compiling images of Earth, the data points are predominantly collated from the US)

3. *Measurement Bias:* data reflects expected correlations often lacking context (e.g. numbers of arrests are used to infer number of expected crimes)
4. *Evaluation Bias:* a one-size-fits-all model used across all categories (e.g. diabetes treatment irrespective of ethnicity and gender)
5. *Aggregation Bias:* when data is aggregated to match a benchmark (e.g. facial recognition algorithms do not work equally accurately on dark-skinned people)

In general, AI systems tend to inherit the biases of the human-driven process that they replace. The more these biases are allowed to exist, the harder they become to inspect or correct. Humans' judgement and biases are easier to reverse if not within one generation then for the next one. Whereas if not corrected, the algorithmic biases become entrenched and may become more difficult to correct than human bias.

According to Professor Joanna Bryson, three main sources of bias have been identified as:[198]

1. Lack of diversity in AI developer teams
2. Lack of adequate testing or flawed selection of data sets for training for machine learning
3. Ill-intended AI developers

To these I would add:

1. *Lack of specific sector deep knowledg*e to understand the real-life implications in a specific sector (The credit decisioning example below shows how this can go all wrong.)
2. *Lack of training for AI developers* in adjacent disciplines like ethics, economics, and humanities like philosophy and theology

Algorithms that affect people's lives should be subject to audit.[199] In this category I have no hesitation about including algorithms used in financial services, primarily those that cater to retail services. Here is an example of a well-seasoned bias framework. An automated credit-decisioning tool using machine learning was released and used for a few years. A number of lenders used it and relied on it as the sole component of their decision-making process.

Naturally, where the system would reject credit applications, it weakened the credit ratings of those applicants. Usually, this decision meant a steep increase in the cost of a loan to the rejected applicant as a result of the reduced credit-rating score. Basically, it pushed people into financial difficulty, by increasing the cost of a loan. A rejected applicant, confused by the decision, given his good score and income, called the lender and asked for an explanation of the decision. The customer service unit provided a blanket answer: "We do not release an explanation as to how our algorithms work. It's proprietary information."Not satisfied with this outcome, the applicant lodged a formal complaint with the regulator, asking for a full disclosure of how the credit decision algorithm worked. It turned out that it was biased to favour social media users. This is a contradiction, as one might argue, "how could someone who spends more time on social media be a better candidate to take a loan?" If one is a diligent, hard-working employee, then surely the amount of time spent on social media would be minimal. How could such a dichotomy be possible? It turns out that the data engineers sourced the data from where it was free and abundant–social media–discarding design consequences derived from a biased data source. It is a fair assumption that bias is present in all AI applications–from facial recognition to credit decisioning. Most of the time such bias is unintended and difficult to spot. However, there are ways to correct such bias. It is rarely easy or simple, but algorithmic auditing needs to be done.

Data strategy needs to have a clear roadmap for how data integrity is checked and safeguarded and bias is removed; ultimately, how fairness in datasets is maintained. The following areas must be considered:

1. Decide what the definition of success is.[200]
2. Accuracy – how often are there errors and for whom does this algorithm fail?
3. Long-term effects of algorithms – data scientists must not be the arbiters of truth but rather the translators[201] of ethical discussions that happen in the larger society.
4. There must be team diversity, with AI developers who are diverse in gender, race, and cultural and economic background.
5. Algorithm influencing must be used in order to weed out bias and correct it. Fact checking has proven a way to influence an algorithm and change its "behaviour."[202]
6. Data sets must be rebalanced by injecting diverse data in the current data training set. The latter is a core research focus at Google.[203]

A group of AI developers advocate for responsible AI licenses, arguing that "developers can include licenses with AI software to restrict its use" to control the potential harmful applications and retain the integrity of the data usage. This is a concept that has been considered as a condition for using open-source datasets like ImageNet (datasets for computer vision applications).

Table 5.1 Companies Developing Tools to Detect Bias

- Accenture has launched the "Fairness Tool," which uses statistical methods to assess AI models. It is focused on that which is "quantifiable and measurable" and aims to help companies assess, identify, and correct biases.
- IBM has Fairness 360 and wants its developers to prove that their algorithms are fair so that users can implicitly trust AI tools.
- Pymetrics, a recruiting start-up, developed an audit AI tool to root out bias in its own algorithms.
- Google has launched the What-If Tool to play with "what if" scenarios of fairness and see the tradeoff.
- WINO Bias focuses on identifying gender bias.
- Wikipedia Toxicity Dataset is a dataset of one million annotations for aggression, toxicity, and personal attacks based on gender, race, and so on.
- Google's Facets and People Research are visualisation tools to help with understanding datasets, and in doing so bias can be visualised.
- Facial recognition datasets have been put together by IBM (36,000 images) and Algorithmic Justice League.

Two thousand nineteen marks the year when the legislators felt that they needed to regulate AI and hold tech companies accountable for their algorithms. In April 2019, US lawmakers introduced a new bill, "Algorithmic Accountability Act," which would require companies to audit their machine-learning systems for bias and discrimination and also correct them when such bias and discrimination is identified. The UK, France, and Australia have also drafted or passed legislation for algorithm accountability.[204] There will be many other countries following this path, and this trend will be embedded in financial services regulations. Since 2016, I have promoted ethical AI in financial services and developed ethics frameworks for AI adoption. Clients have asked why they need to do it in the absence of specific regulation. It's best practice, and when ethics become regulations, it will protect my clients from hefty back-dated regulatory fines for rogue algorithms and complaints that may turn into class-action court cases.

There are many good books written on the perils of data integrity and algorithm design, but there are two in particular that I would recommend. The first is *The Black Box Society: The Secret Algorithms That Control Money and Information* by Frank Pasquale.[205]

It is a thoroughly and thoughtfully researched book, containing first-rate work that combines ideas from different disciplines. It provides a riveting and realistic account of how big players in finance and technology ill-use their power conferred by access to data and powerful algorithms. The second book is *Weapons of Math Destruction: How Big Data Increases Inequality and Threatens Democracy* by Cathy O'Neil. She is a mathematician and former Wall Street trader who provides an insightful yet accessible book founded on a fact-based survey of how the mis-design in mathematical models is destroying trust in our society.

Bias is a business risk comparable to cybersecurity. It needs to be managed and given the right attention by the C-suite and the board. Biased decisions will damage the relationship between the organisation and the public, and as AI continue to make more independent decisions in healthcare, investments, and other financial decisions, the risk of getting it wrong and causing damage to people's lives will increase.[206]

In conclusion, organisations need to be mindful of how data is used, AI design, and AI deployment. It is imperative that the board, innovation teams, and data science teams understand that it is essential that we build and deploy trustworthy AI. When deployed, AI exposes extant problems.[207] AI tools cannot be trusted without a thorough version control attached to them. We need to keep version control for audit purposes and in order to ask the hard questions at the point of procurement as well as deployment.

It is an organisational need to move fast and weave in the business practice of agile ethics right from the procurement stage of an AI tool. Moreover, let's not mistake ethics for compliance. Ethical AI is not compliance–it is good business. It enables the industry to revise and refine their service delivery by putting the customer first. It enables organisations to look at themselves first and ask if they are in the best ethical framework they can be in. This is where redefining trust begins.

PRIVACY

Privacy can be broadly defined as the freedom from interference or intrusion, the right to be left alone.[208] The issues around privacy have, and will, continue to gain momentum and prominence. General Data Protection Regulation (GDPR), a set of rules on data protection and privacy introduced by the European Commission in May 2018, prompted many decision makers to reflect on GDPR's indirect impact globally, beyond the jurisdiction of the EU. In 2019, the EU will further clarify and tighten the rules with ePrivacy regulations, aiming to give users more control over their own data. At the other extreme, China's technology giants are allowed "to gather as much personal data as they like provided the government is granted access."[209]

Privacy will be one of the leading technology themes. Michael Krigsman of CXOTalk conversation summarises some of the core issues to address: "Privacy is not a new concept in computing, the growth of aggregated data magnifies privacy challenges. AI and privacy are both complicit and there are no easy or simple answers because solutions lie at the shifting and conflicted intersection of technology, commercial profit, public policy, and even individual and cultural attitudes." There are two foundational angles to apply to AI and privacy. Both compete and yet hinge on the need for data:

1. In order to build better algorithms, we need more private data, since they are designed for use cases and in the context of creating human experiences. At the same time, in order to de-bias an algorithm that, for instance, favours white males when flagging better-paid job adverts, we need more women' and minorities' data. Obtaining this specific data involves privacy concerns. One way to address it is to work with experts in minorities research, or/and use synthetic data, or anonymise the data. Experts continue to work on privacy engineering. Carnegie Mellon University offers a Master of Science in Information Technology – Privacy Engineering, the first of its kind, in recognition of the need to train experts in this field. Chief Privacy Officer at Cisco, Michelle Dennedy's *The Privacy Engineer's Manifesto* is a good book to explore this topic further. Privacy by design is how privacy engineering manifests in the production stage.
2. Use AI to design new ways to use private data on the edge but not move it from where it is stored, to give people control over their digital lives. Meeco, an Australian company, is one of the pioneers in personal data rights. They have developed a distributed technology powering consent and personal data,[210] and their founder Katryna Dow has a well-articulated vision for access, control, delegation, consent or identity, and personal data that truly puts people first.

In March 2019 the IEEE's Global Initiative on Ethics of Autonomous and Intelligent Systems released a series of useful documents that aim to set valuable frameworks for ethical AI. They are essential reading for the risk-management department of every financial services institution and regulator, and they can be found at *ethicsinaction.ieee.com* and *standards.ieee.org*. I recommend that any management function reporting to the board incorporates references and uses these standards of evaluation. IEEE's set of standards and principles cover a number of topics, and the most relevant to any CRO in financial services might include the following resources:

1. Ethically Aligned Design (EAD) 1st edition, the most comprehensive global treatise regarding the Ethics of Autonomous and Intelligent Systems, which I recommend as the starting point for organisations creating their AI ethics maps. You can find it at *ethicsinaction.ieee.org*
2. Model Process for addressing ethical concerns during systems design
3. Transparency of autonomous systems
4. Data privacy process
5. Algorithmic bias considerations
6. Standards on employer data governance
7. Standards on Personal Data AI agent working group
8. Ontological standards for ethically driven robotics and automation systems
9. Standards for the process of identifying and rating the trustworthiness of news sources
10. Standards for machine readable personal privacy terms
11. Inclusion and application of standards for automated facial analysis technology

The public sentiment towards data privacy is shaping and shifting, with profound discussion surrounding privacy rights for children who have their data harvested either by facial recognition devices as early as kindergarten, school, or regular summer camps where their presence in classes is recorded and their emotions are closely tracked or by smart devices like Amazon's Alexa, Google's AI assistant, or Mattel's toys. Child-tracking devices bought by parents, like the Chinese company MiSafes, which manufactures child-location tracking smartwatches that are sold in the UK, have proven easy to hack. One security researcher discovered that the devices do not encrypt data.[211] Parents need to learn to protect their children's privacy rights and also teach them to protect their privacy rights themselves.

In 2019, facial recognition used by police and other agencies came under close scrutiny, culminating with San Francisco becoming the first major American city to ban the use of this technology by government agencies. Endless ethical discussions around sex robots (sexbots) are fuelled by predictions that 50 percent of Americans think that having sex with a sexbot will become common practice in 50 years and by authorities banning the first "robot" brothel in Houston in 2018.

"It's a real-world challenge that society is about to face for the first time. I hope that the law gets it right," said Francis X. Shen, Associate Professor of Law, University of Minnesota,[212] voicing the concerns of the many. In all of these cases, and many more, the fundamental question is about what happens to this intimate data–who owns it, to what extent can it be used and until when, can you withdraw consent, who uses it, and how does it influence and shape a future adult's life, for instance, their access to credit or jobs?

Privacy International, a London-based charity that defends and promotes the right to privacy across the world, filed complaints against seven leading companies that have a business model based on using private data: Acxiom, Criteo, Equifax, Experian, Oracle, Quantcast, and Tapad, for wide-scale and systematic infringements of data protection laws.[213]

According to a Pew Research Center survey, "Public Attitudes Toward Computer Algorithms," released in November 2018, 57 percent of respondents said they thought it was unacceptable that an algorithm would screen job applicants' resumes, and the majority of the Americans interviewed expressed broad concerns regarding the fairness and effectiveness of algorithms making important decisions in people's lives.

In 2017 the Information Commissioner's Office (ICO) in the UK produced "Big data, artificial intelligence, machine learning and data protection," which also provides specific guidance for privacy impact assessments for big-data analytics projects. They have expressed an interest in the development of some specific guidance on conducting privacy impact assessments (PIAs) in a big-data context. This is highly relevant to AI applications. PIAs are particularly important in this area because of the capabilities of big-data analytics and the potential data-protection implications that can arise. Privacy impact assessments will be required under GDPR, where there might be a risk to the rights and freedoms of individuals. GDPR refers to "data protection impact assessment," which is required in the case of "a systematic and extensive evaluation of personal aspects relating to natural persons which is based on automated processing, including profiling, and on which decisions are based that produce legal effects concerning the natural person or similarly significantly affect the natural person."[214]

Much effort has been exerted in order to explain that anonymised data implies that many concerns about data privacy will be eradicated. That's not the case. A group of researchers aliased three months of credit card records for 1.1 million people. Their work demonstrated that only "four spatiotemporal points are enough to uniquely re-identify 90% of individuals and that knowing the price of a transaction increases the re-identification by 22%, on average." Finally, their work demonstrated that women were more identifiable than men in credit card metadata. [215]

For most AI applications that use large datasets, the data protection impact assessment will be legally required. ICO, a UK organisation set up to uphold information rights, provides some guidance through a code of best practices,[216] which set the scene for the important aspects to consider. This is a valuable checklist to use when formulating your own organisation's internal guidelines. It is equally relevant irrespective of whether you are an AI vendor or an enterprise looking to adopt AI. For an AI vendor, this is useful to build in as version control, as a way to showcase their quality of work and thought. For organisations, this is a useful tool to help evaluate their thinking. In order to get you started, I have listed a summary of the key points. This list is not exhaustive, since its authors regard Privacy Impact Assessment as a living and scalable document. I encourage you to consult this document in its entirety, which is available on *ico.gov.uk*.[217]

In Singapore, the Personal Data Protection Commission produced a discussion paper titled "Artificial Intelligence (AI) and Personal Data – Fostering Responsible Development and Adoption of AI," where their objective is to put forward a preliminary analysis of the key issues essential to the commercial development and adoption of AI solutions. This document is referenced as fundamental in the Monetary Authority of Singapore's *AI in Financial Services: Principles to Promote Fairness, Ethics, Accountability and Transparency* published on 12 November 2018. The first attempts to design and enforce data protection and privacy laws will continue to expand. There is a growing recognition that as data continues to flow around the world, businesses and their boards will likely find themselves spending increasingly more time thinking about the legal risks with transferring and fielding requests to access data internationally, according to Law360.[218] There have been a number of fundamental developments in 2017 that have shifted the narrative and started a snowball effect in the data privacy field. Translating principles into practice is a whole separate discussion. In a well-written paper, "Translating Principles into Practices of Digital Ethics: Five Risks of Being Unethical," Professor Floridi warns of how the digital ethics narrative may be turned into ethics washing–a narrative hollow of meaning but full of mislabelled concepts that sound good but lack the meaning behind the ethics principles.

In an unprecedented move, Tim Cook, the CEO of Apple, in a piece published by *Time* on 17 January 2019, sets the tone for data privacy regulations in the US: "In 2019, it's time to stand up for the right to privacy–yours, mine, all of ours. Consumers shouldn't have to tolerate another year of companies irresponsibly amassing huge user profiles, data breaches that seem out of control and the vanishing ability to control our own digital lives." One key takeaway of this article is that "one of the biggest challenges in protecting privacy [is that] many violations are invisible." Moreover, Tim Cook is calling on the US Federal Trade Commission to establish a "data-broker clearinghouse" that would enable consumers to monitor and demand the deletion of data held by companies that operate under the radar, marking the latest entry in the simmering debate over the best way to regulate tech giants' handling of consumer data. In 2019 at Davos, Microsoft's CEO Satya Nadella called privacy "a human right." In February 2019, Cisco issued a global call to governments to establish privacy as a fundamental human right in the digital economy embedded in updated privacy laws. While for some it may be yet another published article, my readers would be well advised to regard it as a fundamental move of Apple's executive management in positioning Apple as the guardian of user's privacy. I would argue that this is a good business decision, as I believe it will help Apple differentiate itself in the smartphones market and make it the choice for a growing number of privacy-conscious people. This is aligned with Apple's past stances on the privacy of iPhones[219] in 2016, followed by Cook's open criticism of Facebook's mismanagement of private data on the heels of the Cambridge Analytica revelations.

Tim Cook's article signals unequivocally how one of the largest technology firms in the world will lead this discussion. This article suggests the growing momentum of this debate. While client data protection has been hotly guarded and clearly regulated in financial services, it is important for boards and executive teams to focus on how the AI tools they are using are safeguarding data protection. This is yet another reason for boards to equip themselves with the right knowledge and ensure they approve only investments in AI that uphold these values and use future-looking AI technologies.

I would like to conclude this section by quoting Professor Noel Sharkey: "We're welcoming these devices beyond the shores of our privacy like indigenous peoples giving their lands away for glass beads." I invite you to reflect on it.

IDENTITY, AUTHENTICATION, AND AI

As we have become hyperconnected, we have built a *digital identity* for ourselves to represent us in the digital world we engage in so often. "Twenty five years ago, on the Internet, no one knew if you were a dog. Now they do not know if you're a bot controlling a toaster pretending to be a dog," explains Dave Birch, an internationally recognised name in digital identity and author of two books that I recommend: *Identity Is the New Money* and *Before Babylon, Beyond Bitcoin*.

Identity in financial services will continue to remain a core concern, but of a different type. Currently, we see biometrics, such as fingerprints or voice patterns, used primarily in payments, but as the industry is looking to create a seamless experience for their users, either for new clients during the on-boarding process or current clients, we should expect such biometric authentication to permeate other business processes beyond payments. Modern-day biometric authentication technologies have evolved from facial recognition to more sophisticated processes to match higher levels of security like measuring an individual's brainwave activity in response to a particular visual stimuli.[220]

Facial recognition continues to expand as "smile-to-pay" payment systems launched with Alipay technology in more than 300 KFC locations in China. Biometric payments and retail customer interactions are flourishing in China, due to a unique culture and payments infrastructure. In 2018 we saw progress in other types of payment authentication. Fujitsu palm vein biometric authentication for cardless payments has been piloted for Japan's AEON Credit Service. Fujitsu's technology has proven to have high authentication accuracy. Customers register their palm vein biometrics with AEON, and at each point of payment, customers are prompted to input their birth dates and scan their palms.[221] Hitachi has piloted their finger vein biometrics. A Danish debit card firm, Dankort, teamed up with the British biometric identification and authentication service provider Sthaler to create a self-service payment system using Hitachi's VeinID technology.[222] At the end of 2018, Société Générale announced that they had rolled out an improved security system with fingerprint biometrics, looking to remove the amount limit for contactless payments. They are using a biometric bank card issued by IDEMIA, a leader in identity management and authentication, leveraging AI. Singapore announced that they are extending the nation's biometric ID systems to the banking sector.

The Goode Intelligence report shows that a hybrid use in payments, ticketing, identity verification, and age verification is one of the main drivers of a rapid increase in adoption of biometric payments. The same report forecasts that 2.6 billion people globally will use biometrics for payments by 2023, and there will be 579 million cards in use.[223]

The legal profession will be preoccupied with multiple side implications of biometric authentication. For instance, in 2018, a California federal judge ruled that suspects cannot be forced to unlock their devices with biometric identifiers and called such practice unconstitutional. There are approximately 100 class actions alleging violations under Illinois' Biometric Information Privacy Act (BIPA), which regulates the "collection, use, safeguarding, handling, storage, retention, and destruction of biometric identifiers and information" and the compensation when people are aggrieved by the use of biometric tools.[224]

AI is finding a home in biometrics security, especially in biometrics that hinges on behavioural changes like handwriting or typewriting biometrics authentication. This is an AI biometric system that evolves to recognise subtle changes and that becomes more accurate in authentication. In typewriting authentication, an AI biometric system would count for speed, pressure, and hand-dominance. An AI biometric system would identify too many attempts to authenticate as unusual. Biometrics that implement AI have the benefit of continuous monitoring. For instance, in 24/7 monitored rooms or editing confidential documents that need to be accessed by people who have high-level access, the user's typing behaviour can authenticate him or her as the person editing the document and that no data tampering is taking place. AI can also be implemented in more traditional biometric authentication tools, like the iris scan, with the benefit of monitoring insights. More traditional tools can also benefit from computer vision technology to improve their user interface for a faster, less intrusive experience. The next level of advancements is to authenticate the machines that are acting on humans' behalf or just engaging online. The next frontier is to authenticate non-biological entities.[225] BioCatch is one of the best-funded AI companies and provides behavioural authentication and threat-detection solutions for mobile and web applications. BioCatch proactively collects and analyses bio-behavioural, cognitive, and physiological parameters to generate a unique user profile. Banks and mobile apps use BioCatch to significantly reduce friction associated with risky transactions and protect users against cyber threats.[226]

INFORMATION WARFARE

These people have something in common: They do not exist. These faces have been created by generative adversarial networks (GANs), which we talked about in Chapter 2, which generates fake fingerprints that have proven to trick smartphone sensors (all in the academic research stage reported in February 2019).[227] The same GANs generate fake photos[228] of people and anyone can download them from *ThisPersonDoesNotExist.com*.[229] (These photos were downloaded on 19th February 2019). These are deep fakes. As you can see, they are versatile across different age group, gender, face structure, and skin colour. We are in the middle of information warfare and we need to be aware of how this unfolds–and how to protect ourselves and our businesses. What could possibly go wrong in financial services? Everything.

Businesses alike seem ill-equipped to understand the reality of information warfare. In financial services such attacks can aim to acquire data and strategically leak it, and/or wage deceptive public influence or industrial/corporate espionage. Moreover, deepfake videos or photos can be e-mailed as proof of a customer's request to perform a certain activity on his or her account, such as transfering money, closing accounts, or selling stock positions. Such information warfare can also aim to use advanced digital deceptions to distort reality and manipulate the public or customers' opinions and attack a business brand and credibility. These attacks use five main steps to achieve their objective:

1. Reconnaissance – The intruder scrapes and steals the target's metadata.
2. Weaponisation – AI enables editing software used to generate malicious fake content – video, audio, or photos.
3. Attack – Bot armies strategically pump deceptive content online; machine-learning bots feed content to people most likely to share fake media.
4. Infection – Social news and private news feeds enable widespread sharing and viewing of deceptive content.
5. Destruction – Disinformation runs rampant, eroding trust and breeding chaos and confusion; customers close accounts or withdraw money.

A group of Stanford University researchers published research that demonstrates that it is possible "to alter a person's pre-recorded face in real-time to mimic another person's expressions. Essentially, an actor wanting to impersonate a target can create a *digital human puppet* by making a face into a webcam. A digital rendition of the target's face will mimic the actor's face in real-time." Professor Mattias Niesser at niesserlab.org provides further information on this topic.

An area of concern for our industry is that fabricating audio is becoming easier. Lyrebird, a Canadian start-up, is developing technology that can "record one minute of audio from someone's voice to generate longer fabricated audio clips in the same voice."[230] For example, this means that a criminal can record my voice from any of my public talks available online and then fabricate an audio clip to play over the phone to identify themselves as me and access my bank account. How do we protect our customers from such abuses?

A group of computer scientists from New York University's Tandon School of Engineering has developed DeepMasterPrints, a machine-learning tool that generates fake fingerprints that "dupe smartphone sensors but can masquerade as prints from numerous different people."[231] The concept is to combine common fingerprint traits. Given that many banks use fingerprint authentication, biometrics identification tools now need to be hardened against artificial fingerprint attack.

TruePic is a verification platform for videos and photos that uses a watermark on verified files as a result of advanced image forensics using the latest in computer vision, AI, and cryptography technologies to solve the complex task of photo and video verification.[232] They address a host of points that can be manipulated like content, location spoofing, metadata manipulation, rebroadcasting, repurposing existing images, and AI-generated images.

ANTITRUST AND COMPETITION LAW

Typically, access to data is concentrated in the hands of the few. Alongside the design of algorithms deployed in the financial services space, we can see an impact beyond financial products and client engagement. The impact is most profound at the market-structure level. As data-driven technology firms like Google, Amazon, and Ant Financial are making inroads into the financial services space and the Western European space, we need to reflect on their impact on antitrust and competition law. As an industry, there is a need to bring in academic research from centres like the Stigler Center for the Study of Economy and State, at the University of Chicago.

The Stigler Center hosts an annual "Antitrust and Competition Conference."[233] In 2018, they dedicated it to the topic of Digital Platforms and Concentration. One of the speakers was an American antitrust lawyer, Gary Reback of Carr & Ferrell, LLP. His work in trade regulation litigation is well known, and he is "generally credited with spearheading the efforts leading to the US Government's prosecution of Microsoft"[234] in the 1990s. In his contribution[235] at the Stigler Center's conference, he explained the differences between US and EU antitrust regulations, their enforcement, and their efficacy.

Referring to European antitrust officials who fined Google a record $2.7 billion in 2017 for "unfairly favoring some of its own services over those of rivals"[236] and a further €1.5 billion in 2019, Gary Reback[237] noted that the "EU remedy [fines] on Google search [NB algorithm] manipulation case came 10 years *after* Google started their conduct and after they wiped out all their competition. Most of their competitors were already damaged or out of business by then."

He proposes a more efficient way to handle antitrust cases: to put CEOs of companies through court trials, instead of having politicians reviewing their companies' business practices. The latter are yet to prove their impact, as shown by the Facebook and Cambridge Analytica manipulation case when US politicians struggled to pin down Facebook. Gary Reback also referenced the antitrust AT&T case. This is an important example as the AT&T monopoly breakup, paired with pricing policy controlled by the Federal Communications Commission and state regulations, was able to unleash massive innovation, and the development of the internet is mentioned as the most notable consequence.

ECONOMIC ISSUES

As mentioned above, there is a growing concern that data access and control have already created pockets of power and influence concentrated in the hands of the few. There is a questions about whether such concentration will further deepen income inequality. This is the result of not only job displacement but also the majority of people's data being used to the benefit of the few and not of the many. Importantly, the latter are those whose data is harvested and used to develop better and more powerful algorithms and only the few derive monetary benefits from it. We heard it many times but didn't quite internalise it: *If a service is free, then you are the product.* Facebook comes to mind. Experts indicate that this is not uniquely specific to Facebook, as other companies such as Google are often cited. This narrative usually goes hand in hand with privacy concerns.

In 2017, The Future of Life Institute cites Yoshua Bengio, a professor at the University of Montreal, who is known for his founding work in deep learning, saying that in the development of safe AI, the income equality and shared benefits could be pivotal.[238] The economic prosperity created by AI should be shared broadly, to benefit all of humanity. This is the shared prosperity principle, one of 23 issues identified by The Asilomar AI Principles, endorsed and supported by leading names in AI like Elon Musk.[239] IBM committed to a 10-year, $240-million investment to create the MIT–IBM Watson AI Lab in partnership with MIT. One of the focuses of this partnership is to tackle the societal and economic challenges and explore how shared prosperity across nations, enterprises, and people can be achieved.[240]

I believe that shared prosperity through the use of AI might become the top agenda for boards, as a key element of corporate governance and social responsibility.

REGULATIONS, PRINCIPLES, AND GUIDELINES

Some argue that ethical principles and best practices guidance are enough to move forward with the responsible adoption of AI technologies. Their main argument is that legislation will only slow down innovation. This is naive. Our industry has a good chance to redefine trust with AI. We operate in a regulated environment, and so we do not have room for error. We need clear frameworks as we adopt this technology. Where the regulator hasn't provided them, the industry needs to exercise common sense to build their own frameworks. Naturally, this requires specific training tailored to financial services–experts with proven expertise at the intersection of applied AI and financial services.

Others argue that we should regulate only that which has a material impact on people. Again, this is partially applicable to financial services. There is a lot of discussion about the "black box" algorithms that are used primarily in high-frequency trading with no real understanding of how the decision has been reached. I'd argue that depending on the type of algorithm used, it is not a case of "black box" but rather a case of unstructured deployment and no version control. When deciding what to regulate, we are faced with an infinite task. It is important to regulate only what is critical to society, and subsequently requires full explainability. So what do we regulate? Stuart Russell, a well-known professor of computer science at UC Berkeley (US) raises the core principle of *value alignment to human values:* "No one in the field is calling for regulation of basic research. . . . The right response seems to be to change the goals of the field itself; instead of pure intelligence, we need to build intelligence that is *provably* aligned with human values. For practical reasons, we will need to solve the value alignment problem even for relatively unintelligent AI systems that operate in the human environment. There is cause for optimism, if we understand that this issue is an intrinsic part of AI, much as containment is an intrinsic part of modern nuclear fusion research. The world need not be headed for grief."[241]

There seems to be a clear direction from the academic community that guidelines are more desirable than regulations. However, I suspect that it will not be long until the regulators will actually have to step in to normalise some anomalies and possible abuses that have already been identified in AI deployment in financial services. The first step will be light-touch guidelines. For instance, as mentioned earlier in the chapter, in November 2018 the Monetary Authority of Singapore (MAS) issued "principles to promote fairness, ethics, accountability and transparency in the use of Artificial Intelligence and Data Analytics in Singapore's financial sector."[242] This sets a good precedent and incentive for other regulators to support the same light-touch approach. The purpose is to provide some cornerstone principles in implementing AI governance. This regulator has rightly identified that these principles are also intended to "promote public confidence and trust in the use of AI and data analytics."

Notably, in February 2019, the UK government launched a groundbreaking initiative to address the skills gap, a milestone of their modern industrial strategy's AI Sector Deal launched in April 2018. There will be 200 new AI Masters places at UK universities funded by Deep Mind, Cisco, and BAE Systems. The AI Master courses are coupled with work-based placements. There will also be 16 dedicated centres at universities across the country to train the next generation of 1,000 PhD students. University of Bath, part of this initiative, accepts PhD applicants from all education backgrounds, in recognition that AI ethics experts will be required in all lines of work. In March 2019, the UK's House of Lords put forward a clear request to the UK Government for the creation of a Digital Regulator in order to avoid fragmentation of technology and digital regulations in our society. All these initiatives will influence financial services regulations. In March 2019, Stanford University announced that they received $1 billion in endowments to fund the Stanford Institute for Human Centered Artificial Intelligence (Stanford HAI), an interdisciplinary initiative aiming to become the global hub for AI thinkers, researchers, developers, and users across government, policy makers, business, and academia. The degree includes classes from Theoretical Neuroscience, AI in Real Life, the Politics of Algorithms, and Digital Civic Society to Decision Making under Uncertainty, Regulating AI, and Philosophy of AI.

AI Ethics Committee in Businesses

There are a growing number of voices who suggest that businesses that use AI ought to have their own Ethics Advisory Committee. I agree. AI Ethics should be approached in the same way health and safety is approached. Meeting regulatory requirements for designing and deploying this technology should be at the top of the agenda. Deloitte's survey ranked AI ethical issues as one of the top three AI risks, and yet they do not have systems in place to manage any of the AI risks.[243]

Adopters of AI need to scrutinise carefully data sources, design, and uses of AI. It would be wrong to amalgamate AI ethics with compliance requirements. AI systems are self-learning and therefore it's essential that they are designed with care to reflect a company's values, to preempt damages and harm, and to provide transparency when being interrogated. There are many compliance systems vendors that have eagerly jumped on the hyped AI wagon, who generously dispense AI Ethics advice that is glaringly low in quality and insight. AI Ethics is not compliance, regtech, or IT.

AI Ethics requires people who have specialist knowledge in (1) AI technologies, (2) the financial services sector where it is applied, (3) corporate governance, and (4) AI ethics.

An AI Ethics committee should be part of the technology board and report to the board of directors. Its role would be to manage the specific risks of using AI technologies: vet AI systems, advise on AI strategy, propose remedial actions, and ensure that the company has a joined up AI ethics strategy for self-regulation. When done correctly, such self-regulation will enable companies to use AI that truly puts customers first, one of the core principles of The Financial Conduct Authority's code of conduct for financial services.

Black Box No More: Explainable AI

AI specialists argue that there is a drawback in aiming for full explainability, because the accuracy is indirectly correlated with explainability. There is work to be done to reduce this correlation. I am not advocating delaying it, but in the meantime we need to be realistic about the importance of accuracy and explainability on a process-by-process basis.

The need to provide explainable AI tools in financial services has been taken seriously by AI vendors like Fujitsu. Building explainable tools is expensive but not impossible, as some have argued. Fujitsu Laboratories announced their unique technologies for explainable AI. Their solution is about connecting two technologies: Deep Tensor® and Knowledge Graph. Deep Tensor is a unique technology based on machine learning, while Knowledge Graph is a knowledge base that presents graph-structured data obtained from documents and databases. Fujitsu has combined the two technologies to help solve the problem of black box AI.[244] The Fujitsu solution is a completely new approach that gives two explanations: (1) reasons for findings and (2) basis for findings.

Fujitsu's explainable AI (diagram below *Source Fujitsu Laboratories; reproduced with permission)* has already been used in credit risk assessment for corporate loans. It has proven a valuable tool to assess the creditworthiness of small to medium size companies that do not have a substantial history–or even lack balance sheets. Fujitsu's tool assesses creditworthiness based on bank-transaction data. Fujitsu sees their explainable AI technologies as useful to other processes in the financial services sector like credit risk assessment for retail finance, prevention of money laundering, and marketing initiatives.

Figure 5.2 Explainable AI Structure

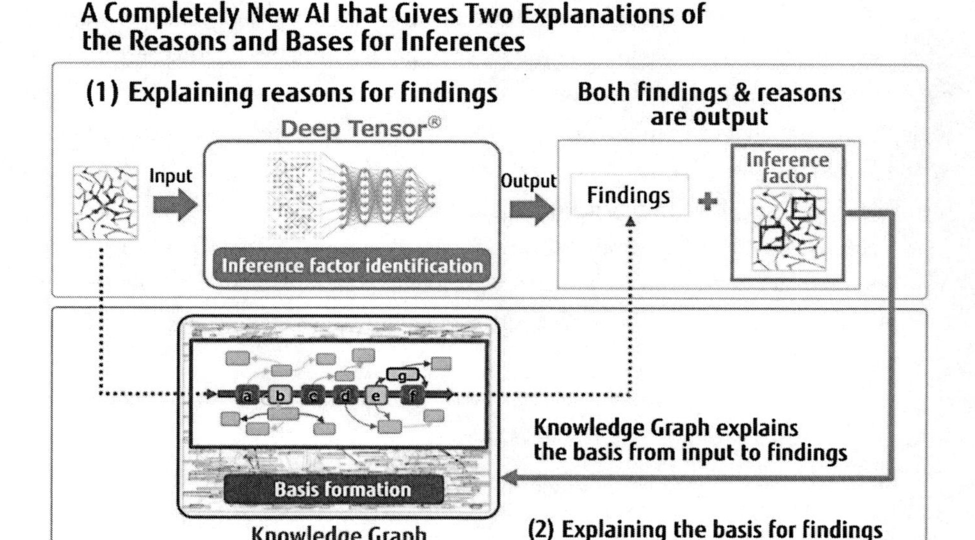

The space of explainable AI is slowly starting to become populated with independent work and the exploration of methods to make machine learning explainable. One example is the book *Interpretable Machine Learning,* a guide for making black box models explainable. Written by Christoph Molnar, a German statistician and machine-learning specialist, the book talks about the interpretation methods and explains how they work, their strengths and weaknesses, and how their outputs can be interpreted. The book focuses on machine-learning models for structured data or tabular data and less on computer vision and natural language processing tasks.

Algorithmic Collusion

Deep-learning algorithms have proven to be to more pliable than machine-learning ones in addressing with greater accuracy (1) overfitting and (2) volatility in the market that manifests as a non-linear event. Deep learning has been adopted at an exponential rate, with Google leading with over 100 projects in 2017 from only two projects five years earlier.[245] This exponential adoption demonstrated many revolutionary advancements. Google's cofounder Sergey Brin admitted openly that "The revolution in deep learning has been very profound. It definitely surprised me, even though I was sitting right here."

However, despite its success in other applications like search and personalised advertising, in trading, verification is key to securing good governance of AI. There are a few examples when unverified trading algorithms were deployed that demonstrate how costly it is to deploy unverified AI algorithms (see chart below). For further reading I recommend the World Economic Forum's essential research on AI principles published in 2019, focusing on financial services, also covering algorithmic collusion.

The Turing Institute published a paper by Professor Bonnie G. Buchanan that provides a useful summary of well-known algorithmic collusions and incidents that also happened due to lack of thorough verification before going live in the market.

Table 5.2 Well-known Lack of Verification Events

YEAR	INSTITUTION/ EVENT	COMMENTS
2012	Knight Capital	$440 million loss in 45 minutes after deploying the trading algorithms
2010	The Flash Crash	P&G stock was traded between 1 penny and $100,000. Problem caused by bugs and computer malfunction
2013	Goldman Sachs	Incorrectly launched 800,000 contracts linked to equities and ETFs
2013	Everbright Securities	Malfunction in algorithms and purchased shares valued at $4 billion(sic) on the Shanghai stock exchange
2016	Betterment LLC	After Brexit referendum, the market volatility was at record high and Betterment relies on algorithmic trading and suspended trading to avoid excessive trading costs for their private clients' portfolios.

THE GEOPOLITICS OF A UNIFIED AI GOVERNANCE

In the past five years, AI global growth has reached exponential levels, fuelled by unprecedented funding, originating from governments and capital markets–private and public alike. These investments are committed to a future that we haven't experienced before. However, the certainty of some investors is worth noting. The Japanese Government

Pension Investment Fund, the world's biggest manager of retirement savings, have openly expressed their view that AI will ultimately replace human fund managers.

AI-first companies have moved from being mere companies with commercial interests to being important foreign actors, performing oversight and regulatory functions and affecting global dynamics. In 2019, Denmark appointed a civil servant as the country's technology ambassador to Silicon Valley. This is a first in the world, and similar appointments should be expected.

As early as June 2016, Japan's Ministry of Internal Affairs and Communication[246] proposed nine principles for developers of AI, which were then submitted for discussion to the G7 summit, which includes the EU, and to the Organisation for Economic Co-operation and Development (OECD). Japan has made it clear that their proposals should be regarded as a soft law, rather than hard regulations. Their initiative is well guided and visionary and has proposed the following nine principles:

1. Collaboration – interconnectivity and interoperability of AI systems
2. Transparency – verifiability of inputs/outputs of AI systems and of their judgements
3. Controllability – pay attention to controllability of these systems
4. Safety – AI systems mustn't hurt human life, body, property of users, or third parties
5. Security – pay attention to security of these systems
6. Privacy – AI systems should not infringe on the privacy of users and third parties
7. Ethics – respect human dignity and individual autonomy in R&D of AI systems
8. User Assistance – AI systems must be designed to support users and make it possible to give them opportunities for choice in appropriate manners
9. Accountability – AI systems should be accountable to stakeholders, including users of AI systems

I'd argue that these principles are also relevant to AI design and deployment in financial services. If we truly want to use this technology in our favour as an industry, we need to understand that its design must be regarded as a global rather than country-based pursuit. We need to build one single language of conversation for AI, rather than a Babel Tower of a multitude of languages and principles that cannot connect and correlate. *The common denominator is to use AI for good.* This value cannot be unlocked if there is no common understanding of what this means and how this is made possible. This can be accomplished through collaborative global governance. Many see that one immediate value of such common language would be the identification and prevention of the 2008 type of financial crises. For such a pursuit alone, it is worth considering a global approach to designing and deployment of AI in financial services. Amy Webb, the author of *The Big Nine* (a book that analyses the top nine AI companies in the world), suggests that an international oversight board with enforcement powers like The International Atomic Energy Agency, which polices countries' nuclear programs, is also needed for AI global oversight.

There are many initiatives run by think tanks like the World Economic Forum or governments to design their own country-level AI strategy. This is a good thing. The financial services' regulators and central banks should be involved and pay close attention to how the government-level thinking is shaping up. These country-level initiatives seem to involve a disproportionate number of academics, leaving our industry with little or no say in the debate. While it is essential to draw from academic thinking, ultimately, AI adoption is not an academic exercise run in isolation, but one that will involve execution with real-life consequences. With that in mind, I urge AI initiatives that are run by governments to involve industry-specific experts. I would also urge the leading think tanks like The World Economic Forum to try to be inclusive when drafting AI global initiatives instead of continuing to involve a small circle of stratospheric business personalities who live an isolated life in relation to the rest of their respective industry. Ethical AI also means inclusion. Ultimately, AI means actionable insights to safeguard the wealth and well-being of nations and individuals. To secure this level of insight, we need to devise a globally collaborative approach to AI strategy in financial services. This collaborative spirit is needed when designing AI strategies. It has to be the common thread of AI initiatives inside the corporation, at the industry level, at the governmental level, and finally at the global level.

In order to safeguard trust and the positive impact this technology is likely to have in financial services, a measured approach to regulating AI might be needed. Earlier in the chapter I explained governing frameworks for regulating AI. It is also essential that the regulators have a strong grasp of how AI technologies work. This is essential because in order to regulate you need to understand what you regulate and *what you say yes to,* as Kay Firth-Butterfield of World Economic Forum aptly summarised it.

MANAGING AI RISKS

Technology is fragile, and its adoption should be examined as a whole and all implications considered, especially those on humans. Technology never eliminates risk, rather it redefines and relocates risk. Corporations should absorb the cost of developing AI safety because, like social corporate responsibility, it is necessary and actually yields good business in the long term. Should managing AI risks be yet another objective for risk management teams? Perhaps, yes, would be my immediate answer. How do we do it? Where do we start? To complement the traditional risk management and data science teams, here is an example of culture change recommendations (even though some may regard them as touchy feely, they are not):

- Build a reservoir of human-centric resources–people who share these values.
- Educate company leadership to have a grasp of AI, as well as humanities.
- Bring in employees with humanist backgrounds like philosophy and theology.
- Promote high moral values like integrity and compassion.
- Imbue your internal policies and staff assessments to evaluate actions to recognise and reward when people act with integrity and trustworthiness.
- Explain these concepts through internal training.
- Make trust your organisation's core value.
- Build strong ties in AI project delivery between business, data science teams, and experts with training in anthropology, history, or theology.
- Set internal regulations to promote human-centric values.

"We need technologists who understand history, who understand economics, who are in conversations with philosophers," said Marina Gorbis, executive director of the Institute for the Future. "Those conversations are essential to ensuring AI does more good than harm."[247] Too true. These technologists design our lives, our business, and our future.

Artificial General Intelligence (AGI)

The topic of AGI, the super-human intelligence, has captured the media's imagination, garnished with Hollywood movie references to *The Terminator*. We have seen a host of science-fiction scenarios described and assertions made that quite possibly have no roots in technological possibilities. One persistent assertion was that AI will expand from its narrow state and reach singularity–a state where it will be smarter than human–and it will bring about the end of the human race.

This is just a hypothesis. As far as the end of the human race, Michael Wooldridge, head of the department of computer science at the University of Oxford, explains it rather clearly "for the foreseeable future, it is a safe bet that far more people will die from natural stupidity than from artificial intelligence."[248]

AGI is described to surpass human processing capabilities and optimise how it operates across all disciplines and industries when it reaches singularity. The assertion is that there are a large number of algorithms collecting and accessing information across all sectors. They are building understanding of human knowledge at a very high pace (Wikipedia is one source) and the individual's preferences via Google search or Facebook or shopping preferences on Amazon. It is possible that if someone were to aggregate the distributed knowledge processing capabilities of these algorithms, we could argue that they would in the future form an AGI, an omniscient tool with which humans are not biologically developing fast enough to be on equal footing. Today, tech firms build world changing technology that influences, permeates and learns about how we our lives, how we construct our social interactions. This is already altering our society in profound ways, which we couldn't predict 5 or 10 years ago. AI technologies have had knock-on effects, some of which are irreversible. Experts indicate that there is a risk of a "runaway reaction" associated with deployment of these technologies where computers' intelligence reaches and exceeds human collective intelligence at a speed that is exponential, so its effects will show in an extremely short period of time. At that moment, the effects, experts argue, will be less about the socio-economic impact on our societies an on our mental health, and more about how these hyper intelligent computers will have their own goals, which they would achieve my manipulating humans' volition and decision making. This is singularity. Is this scenario even possible? Is there anyone who started the journey to amass this vast volume of data points about us, about how we live, and about our society? Some people argue that two tech companies have already started this data collection. In 2019, there are two main big tech companies that are heavily funded, with fast growing cloud platforms, with a large enterprise footprint and their own AI hardware and platform for developers. They are both polarising the move to bring about AGI: (1) Google with their Google AI, and (2) Microsoft with their partnership with Open AI. Their strategic acquisition and partnership programs are also helping them become likely contenders to develop AGI. For instance, in November 2019, Google announced the acquisition of FitBit, the wearable company, which has built one of the largest human health databases. This acquisition will have many implications; one being a likely impact in life insurance premiums calculation. Google's access to personal data including human health data will enable them to build an unprecedented level of knowledge and understanding.

Some say that AGI will not happen for a long time, therefore we do not need to worry about it now. Irrespective of whether singularity may happen or not, I would argue that we actually need to get decidedly concerned *now* for what this unchartered *future might bring.*

EXAMPLES OF QUESTIONS FOR BOARDS TO ASK

	TOPIC	QUESTIONS
1	Internal Proceses	1. What are our ethical AI internal principles and procedures? Who is in charge of designing them and monitoring their implementation? 2. Do we have an AI ethics expert appointed on our technology board advisory committee? Have we considered a board observer seat for such an expert? 3. Have we discussed creating an AI Ethics Officer or Chief Value Officer role? If no, why not?
2	Recovery	1. Do we have any disaster recovery plans if our data is leaked or lost, if someone takes control over our AI systems, or if there is no internet connection? 2. What mechanisms do we have in place to protect our customers and their data and restore trust with them? What's our business disaster recover strategy? 3. What contingency plans do we have in case of a data breach? Where is our data stored (jurisdiction of the cloud provider)?
3	Explainability	1. Are any of our AI tools doing anything that we cannot explain and that is important for our users to know? 2. If yes, what risks are we facing if this were to break out in the news?
4	Human-centric AI	1.What systems do we need to upgrade so that the board ensures that our company improves continuously to become a human-centric value and responsible organisation? 2. Are there any indirect stakeholders that our AI tools impact–people, processes, systems, society at large? 3. Could any of our AI tools have a malicious impact on children, seniors, the disempowered, etc.? Would you allow your children or parents to use it?
5	Accessibility and Diversity	1. Will a diverse set of users be able to access our services? Have we made provisions for special needs people? 2. Can we clarify the diversity in our training dataset and ensure we have the diverse monitoring team in place to identify and correct the bias in the source of that data and design of our algorithms?

	TOPIC	QUESTIONS
5	Bias	1.Do we have mechanisms in place to test the fairness, accountability, truthfulness, and transparency of any algorithms embedded in our product or service right at the development stage? 2. Have we tested our product for bias with users of a different age, gender, race, socioeconomic status and income, geography, political affiliation, language, ability, sexual orientation, religion, and education?
6	AI Risks	1. Have we assessed how a bad actor might hack into our product (for instance, artificial fingerprint attack to clients' bank accounts)? 2. What is our strategy to prioritise the AI risks? What unintended consequences might we face? 3. Which are the biggest AI risks? Which are the most difficult to address? Why? What strategies can help mitigate our AI risks? 4. What is the probability that our AI tools can be hacked and be used on our behalf? How can we take control if that were to happen?
7	AI Governance	1. Where and how should we begin with governance of AI? 2. Who else needs to join our board before we can move forward with redesigning our business strategy with AI? 3. What plans do we have for a technology advisory group to support the board? Can we bring onto our board an AI expert who has our sector business knowledge ?

KEY POINTS TO REMEMBER

- The governance of AI rests on four core intertwined values: (1) AI needs to have a human-centric design, (2) AI design has to promote positive values, (3) AI needs to have a switch-off button, and (4) AI needs to be explainable and the humans who build it and deploy it are accountable.
- The five important facets of AI governance are: opportunities, unintended consequences, philosophical implications, challenges, and academic research in AI governance.
- The study of ethics for AI programmers and researchers might need to become mandatory.

- We must be careful of using data to manipulate people's choices. The danger of digital assistants is when humans are not informed that they are talking to a machine.
- The seven key requirements that trustworthy AI systems should have to make them human-centric include: human agency and oversight, technical robustness and safety, privacy and data governance, transparency, diversity/non-discrimination/fairness, societal and environmental well-being, and accountability.
- Interpretability, transparency, and explainability are all important as an aggregate, as well as independently. They all carry different value propositions for the task at hand. We need to have clarity of purpose and common sense in order to decide when explainability is actually needed. AI safety is an ongoing concern that needs to be addressed. We must not wait until human-level artificial intelligence is achieved–we should start now.
- Algorithmic bias can occur when it is hard coded by the developer, when it comes from biased data, when it is derived from a biased choice of features analysed and how we frame the problem to solve, or from overfitting.
- To reduce bias we need AI developers who are diverse in gender, race, and cultural and economic background.
- While client data protection has been hotly guarded and clearly regulated in financial services, it is important for boards and executive teams to focus on how the AI tools they are using are safeguarding data protection.
- We are in the middle of information warfare and we need to be aware of how this unfolds –and how to protect ourselves and our businesses. In financial services such attacks can aim to acquire data and strategically leak it, and/or wage deceptive public influence or industrial/corporate espionage. Moreover, deepfake videos or photos can be e-mailed as proof of a customer's request to perform a certain activity on his or her account, such as transferring money, closing accounts, or selling stock positions.
- Shared prosperity through the use of AI might become the top agenda for boards, as a key element of corporate governance and social responsibility.
- We need clear frameworks as we adopt this technology. Where the regulator hasn't provided them, the industry needs to exercise common sense to build their own frameworks.
- If we truly want to use this technology in our favour as an industry, we need to understand that its design must be regarded as a global rather than country-based pursuit. *The common denominator is to use AI for good.*
- Managing AI risks should be another objective for risk management teams and needs to be on boards' agenda.

CHAPTER SIX

Boardroom Oversight and AI Adoption

According to the UK's Financial Reporting Council, the board's role is "to provide entrepreneurial leadership of the company within a framework of prudent and effective controls which enables risk to be assessed and managed." In order to do so, I believe that boards must educate themselves about AI and its role in their business. Boards need to have a minimum knowledge to understand how AI is transforming their organisation's competitive landscape and what tools they have at their disposal to drive a long-term growth strategy.

This chapter aims to take a closer look at ways to reframe boards' strategic thinking on AI adoption. How unprepared are the majority of leaders and boards for the digital transformation? Are they comfortable with the technology and do they understand how it works? Furthermore, AI is a business strategy issue not just an IT issue–it needs a blueprint for the correct adoption of AI strategy and the right budget to match it.

Boards would be well advised to avoid signing off a piecemeal adoption of AI. They need to ask executives for a long-term (10-year) roadmap for business growth. Joined up transformation is key to correct AI adoption. AI adoption and AI governance fall within the responsibilities of the boards, which should ask the executives if the organisation is doing enough to maximise the opportunities AI offers and to deploy AI ethically.

Boards of directors often lack insight into the implications of the changing nature of skill sets and training that the workforce will need in the future, as AI enables the automation of current roles. In addition, modern corporate governance needs a closer analysis and focus on improvement in risk and change-management strategies. This includes managing cybersecurity challenges and managing ethical risks. AI and other emerging technologies pose a new range of disruptive risks that require a new approach.

Boards have a variety of challenges that keep growing in complexity and inter-dependability. The London based Better Boards[249], is a specialist data-driven advisory for boards. They identified a useful framework to follow the current challenges:

1. Increasing emphasis on governance
2. Growing concern for environmental and societal issues
3. Increasing importance of stakeholders
4. Growing influence of fund managers/activist investors
5. Increasing distrust of external auditors
6. Increasing demand on corporate directors
7. Changing board composition

The modern board has a multitude of responsibilities to execute on: risk management, oversight, and business growth. One US board director told me that it is not their board's responsibility to design and pursue business strategy. That line of thinking would be devastating for any company to follow. According to BCG, the consulting group, nearly 66 percent of 100 companies with the largest stock price drops from 1995 to 2004 found themselves in this situation not because the financial landscape changed, but because their competitive strength, their operations, and their strategy failed them.[250] McKinsey found that corporate boards discuss business strategy only once a year. This is not enough, considering the speed of change and how digital is turning the whole business landscape upside down.[251] The Spencer Stuart 2018 annual survey of 230 board directors identified one overarching worry they all share: that they are unable to keep up with the correct pace of change and technological innovation. The National Association of Corporate Directors and Partners (NACD) in their 2019 Governance Outlook (projections on emerging board matters) have identified that:

- 70 percent of the public corporate directors surveyed reported that their boards need to strengthen their understanding of the risks and opportunities affecting company performance.
- 62 percent view "atypical or disruptive risks as much more important to the organisation today than five years ago."
- 47 percent see AI as "the biggest technology disruptor."
- 49 percent regard AI as "the biggest business enabler."
- 50 percent of the board plan to improve their oversight of digital transformation.

According to ICSA, the UK's Governance Institute of the FTSE 350 UK Boards, the board directors are focusing their attention on technological disruption. This item has moved higher up the agenda, with 48 percent of boards now having *discussed* the risks and opportunities of Automated Intelligence as opposed to just 36 percent in a 2017 survey.[252] According to Gartner's 2019 CIO (Chief Information Officer) survey, these numbers have shifted rather importantly across all industries. In organisations that were AI early adopters, IT departments are the primary driver of projects compared to two years ago when the C-suite initiated two-thirds of AI initiatives. There are expected challenges facing the leaders and laggards in AI adoption: Laggards struggle to gain leadership support and to define use cases, whereas leaders have shifted the discourse from *IF we can do this* to *HOW we can do it.* Later in the chapter, I'll introduce you to a useful framework called Technology Business Management (TBM) necessary for CIOs (Chief Information Offices) to manage technology infrastructure and to prioritise investments.

Table 6.1 The Technology Issues Modern Boards Are Facing

Issues on the forefront of board's minds

1. AI
2. Big Data
3. Internet of Things (IOT)
4. Cloud
5. Robotics
6. Process Automation

Underlying issues that do not receive necessary focus

1. Connectivity between new technologies
2. Current and future talent needs
3. Cyber-risk mitigation
4. Agile change-ready culture
5. Appropriate incentives and time horizons
6. Internal collaboration IT and business
7. Short- and long-term success metrics
8. Understanding of current and future competition
9. Effective compliance management
10. Business model innovation
11. Optimal capital and resource allocation

AI business-related impact is expanding exponentially. AI has an influence on how big data is used for real-time internal auditing and decision making at the board level.

BOARD OVERSIGHT OF AI

As business models cross over into the data-driven environment, where digitalisation of business models is becoming the norm, business models tend to be regarded more as IT issues rather than board issues. This is wrong in scope and responsibility. IT has grown from a mere operational infrastructure to be the core of business risk as well as strategy deployment and design. It is good corporate governance to regard IT, and any digital and AI-related developments, including AI ethics, as a board issue.

Chapter 5 details a range of governance concerns that the boards needs to be aware of. Reputational risk needs to be managed constantly and carefully. For example, in the machine-learning company that developed COMPAS for parole recommendation for US courts, granular scrutiny of its algorithm has been identified as having provided racially oriented bias in its recommendations. A credit-rating assessment tool has been in use for over four years and has recently been identified with a design flaw that incorrectly favoured those applicants who use social media. Reputational crisis can bring down a company. Therefore, AI ethics is a board issue, and it needs to be addressed carefully and thoughtfully as part of a joined-up AI strategy which I talk about in greater detail in Chapter 7. Perhaps the most constructive way for boards to address AI ethics would be to have detailed guidelines and insights from their technology committee.[253]

In their report "Is Board Risk Oversight Addressing the Right Risks? Strategies for Addressing the New Risk Landscape," the NACD (National Association of Corporate Directors) recommend themes that are relevant board practices and that would prepare board directors to deliver improved oversight of technological disruption in general (reproduced below with permission):

Ensure Management Integrates Disruption Considerations

Company exposure to disruptive change presents a choice: On which side of the change curve do organisations want to be? For example, organisations need to make a conscious decision about whether they are going to be the disrupter and try to lead as a transformer of the industry or, alter- natively, whether they are going to play a waiting game, monitor the competitive landscape, and react appropriately–and in a timely manner–to defend their market share. It is important that the board ground its disruptive-risk oversight with a solid understanding of the company's key strategic drivers and of the significant assumptions made by management that underpin the strategy. Boards should ask management whether they:

1. monitor significant risks related to the execution of the strategy and business model and consider the enterprise's risk appetite and risk tolerances in meeting key objectives;
2. evaluate the risk-reward balance associated with different strategic alternatives to understand the risks the enterprise is taking on as a result of each alternative for creating enterprise value;
3. track the external environment and macroeconomic trends for changes in significant assumptions underlying the strategy and continued relevance of the business model, and evaluate whether disruptive trends exacerbate risk or create market opportunities;
4. integrate lead indicators and advanced data analytics into performance monitoring so that it becomes more anticipatory and forward-looking and supports risk-informed decision making and increased accountability; and
5. involve the board in key decisions e.g., acquisitions of new businesses that could offer access to disruptive technologies, entry into new markets, digital transformation initiatives, or alterations of key assumptions underlying the strategy and invite challenge and open discussion regarding those decisions.[254]

Assess the Continued Effectiveness of the Risk-management Program

Given the pace of change experienced in the industry and the nature and relative riskiness of the organisation's operations, does the board understand the quality of the enterprise risk management (ERM) process informing its risk oversight? How actionable is management's risk information for decision making?

Does ERM effectively capture and assess early warning signals that indicate more unusual or disruptive risks on the horizon? These and other questions focus on the robustness and maturity of the risk-management process. Directors should ensure that the critical attributes of risk-oversight excellence are present:

1. Critical and potentially disruptive enterprise risks are differentiated from the day-to-day risks of managing the business so as to focus the dialogue on the risks that matter to the C-suite and the board.

2. Accountability is established for both traditional and disruptive risks and clearly embedded in the lines of business and core processes.
3. Actionable new risk information is not only reported up but also widely shared to enable more informed decision making.
4. An open, positive dialogue for identifying and evaluating opportunities and risks is encouraged. Consideration should be given to reducing the risk of undue bias and groupthink so that adequate attention is paid to differences in viewpoints that may exist among different executives.
5. Advancements in the application of new technologies–including AI, mobile technologies, advanced data analytics, and visualisation techniques–are used by the organisation to strengthen risk prevention, detection, and mitigation.[255]

Improve the Visibility of Disruptive Risks

In an NACD poll earlier this year, 53 percent of directors cited the lack of information from management as a key barrier that either somewhat or to a great extent hindered effective oversight of atypical, disruptive risks. The 2018 Blue Ribbon Commission report highlights the fact that lots of valuable information about exposure to disruptive risks already exists within companies, but it does not always reach senior management and the board on time.

The report recommends a number of concrete steps to improve management's reporting to the board:

1. Better leverage the internal audit team to share insights about potentially disruptive risks. They possess a wealth of independent information about possible exposures and red flags.
2. Periodically review the format and content of risk reports to ensure they provide sufficiently forward-looking views of potential risks, including new patterns and linkages between different types of risks.
3. Ensure management reporting considers independent, external data about the company's risk profile and evolving environment.
4. Frequently evaluate the current protocols for escalating information to the board. Do processes established to ensure the proper and timely flow of information to the board keep pace with changes in the business and risk environment? Are reporting thresholds clearly established and well understood?[256]

Invest in the Skills Necessary to Successfully Navigate Disruptive Risks

Directors express doubts about the readiness of their own boards to provide effective oversight of disruptive risks: 74 percent of respondents to NACD's online poll held in 2018 reported that lack of board knowledge hinders oversight of disruptive risk to at least some extent.[257]

Table 6.2 The Core IT Infrastructure Focus Points

Below are a range of focus points for boards and executives to monitor continuously as ongoing concerns at the board level:

External environment within which the organisation operates:

1. infrastructure providers (cloud or external servers)
2. software and product development
3. outsourcing of IT services
4. cyberterrorism
5. regulatory requirements
6. cybercrime
7. vendor management, including procurement
8. money laundering
9. fraud prevention and management

Internal environment needs to be centred on people and culture:

1. legacy issues and how to overcome them
2. operational resilience and monitoring
3. platform delivery
4. network architecture
5. software/product development
6. strategic considerations and business redesign

The 2018 Blue Ribbon Commission report outlines a number of action steps:

1. Ensure that the selection and evaluation criteria for the CEO and other senior leaders focus on disruption and resiliency, including success in the following areas:
    - Leading the development of ideas and insights about future trends and opportunities
    - Problem-solving and executing successfully in uncertain situations
    - Openness to alternative points of view and early-stage ideas as well as willingness to question assumptions (one's own, and those of others)[258]
2. Strengthen board oversight of the talent strategy by discussing how disruptive risks factor into the organisation's human capital plans and whether leadership development, compensation, and reward systems reflect the realities of a rapidly changing operating environment.[259]
3. Establish requirements for ongoing learning by all directors and incorporate them into the board-evaluation process. Directors need to invest in continuous learning and development in order to grasp the company-specific impact of disruptive new risks and opportunities and to maintain an independent, well-informed point of view about the business and industry. Nominating and governance committees should ask directors to provide updates about how they are taking a proactive approach to ongoing learning.

Key Risks and Legal Issues

In 2017, the Financial Stability Board (FSB), based in Basel, Switzerland, issued a revised survey of benefits and risks associated with the growing use of AI for financial stability of the sector. They took a balanced view and analysed the positives as well as potential risks of AI deployment.

Their findings and recommendations around the governance of AI make a good starting point for corporate boards to reflect on potential risks. I have listed below a few key risks and key legal issues for boards to continuously monitor:

Key Risks:

1. Network effects and scalability of new technologies may give rise to third-party dependencies. This could in turn lead to the emergence of new systemically important players that might fall outside of regulatory oversight.
2. Interpretability and auditability issues, which, if not addressed, could present macro-level risks for the financial system.

3. The use of "black box" or opaque models, which could result in unforeseen and unintended consequences.
4. Maintaining IT governance and information security at the highest levels–they are as important as capital and liquidity requirements.
5. Implement effective control environments necessary to support innovation, including appropriate processes for due diligence, risk assessment, and ongoing monitoring of any operations outsourced to a third party.
6. Ongoing risk monitoring of and a robust development process of machine-learning models. In the European Union, this includes the requirement of continuous monitoring of risks associated with adoption of AI.

Key Legal Issues:
1. Data privacy and protection, consumer protection, anti-discrimination, and liability issues
2. Allocation of responsibility amongst AI vendors, operators, and users of AI
3. Liability issues, the FSB emphasises, will become increasingly important, given the possibility of mistakes and legal disputes around damages

The FSB are cautiously optimistic and recommend that the use of AI should be monitored continuously and that risks need to be considered at each stage. Finally, they also recommend that continuous monitoring of innovations in the future should remain on the boards's agenda.[260]

BOARD PRODUCTIVITY WITH AI

Board meetings are not always as productive or efficient as they could be. McKinsey reported that directors spend 70 percent of their meetings on quarterly reports, audit reviews, compliance, and financial aspects like budgeting, leaving very little time to discuss the strategy and direction of the business.[261] Board members struggle with a lack of time, onerous reading material in board packs, and the challenge of making decisions under uncertainty. AI tools offer notable new developments that can save time, make meetings more effective, and provide real-time data for decision making.

Diligent, a US company, offers a noteworthy board-management software. Their work is essential to managing enterprise corporate governance. In terms of technological advancement, what caught my attention with their work was that they have recently launched their own governance cloud. This offering helps directors streamline duties for compliance, regulation, and governance while keeping all processes in a highly secure, confidential platform,[262] and I can imagine that with the right AI tools, it will help Diligent scale while remaining highly secure and independent of the usual cloud providers.

Board Packs

The public scrutiny of board materials ranges from regulatory to parliamentary. Despite this, in the UK, nearly two out of three company secretaries rate their board pack as "weak" or "poor." And yet no substantial improvement is being achieved. The most cited reason is the lack of time. Board directors struggle with information overload. A board pack is on average 500 pages (about the size of a book and a half). This stretches the board directors to impossible levels to read, understand, and then make decisions. The problem is compounded by the fact that the majority of board directors are non-executive directors with a portfolio career serving on multiple boards. A board meets about six times a year, and some non-executive directors (NED) sit on about four different boards. This means that an NED reads on average about 36 books a year in board packs alone. The volume is staggering, but the essential component is the liability the directors have for remembering and applying correctly the information they receive from board packs.

The ICAEW's (The Institute of Chartered Accountants in England and Wales) report commission by the Prudential Regulation Authority, a UK regulatory body in financial services, found that sometimes board packs can be 1,000 pages long. The risk of missing out on crucial details is real. The UK has introduced the Senior Manager and Certification regime in financial services. This regulatory requirement has placed yet another layer of complexity, volume of information, and liability on board directors. This report has indicated that 92 percent of board packs now contain more regulatory and compliance reporting, 88 percent of board packs now contain more risk reporting, and that time spent on reading board packs has increased by 30 percent since 2011.[263]

There are companies like Board Intelligence that focus on preparing and managing board packs. Board Intelligence prepares a large majority of board packs in the UK. They have a trained team of specialists who summarise thousands of pages of a board pack into approximately 500 pages. The document management service of board packs is done online, via an interface that enables visualisation and distributes an electronic version of these documents to the rest of board. Organisations should be able to purchase or build their own dedicated AI tool for the board, which would enable their board meetings, packs, and action points to be managed internally, which means confidentially secured. Progress can be achieved in reducing the time and increasing the efficiency of board documents and meetings in a very short space of time. As discussed earlier in the book, in 2018, deep-learning technologies have delivered tasks like document analysis, reading comprehension, and document summarisation, which have reached human parity in terms of accuracy. These are extremely useful developments for boards in financial services, and boards should make use of these developments. Document summarisation, content extraction, searching, and document indexing are all tasks that can be achieved with dedicated Natural Language Processing, a subset of AI technologies (discussed in Chapter 2: The taxonomy of AI).

Decision Making

Board experts have recently discussed the extent to which "human directors should be allowed or required to rely on artificial intelligence" to make their decisions.[264] JP Morgan's board has access to real-time data and reports, which can be produced shortly before or during board meetings. It might make sense to rely on real-time data in order to make the best decisions. Salesforce's CEO Marc Benioff has a personalised version of Einstein, their AI tool, which helps him run his company.

Decision making under uncertainty has been one of the core interests in applying machine learning to resource-constrained scheduling problems. Work in this field has been conducted for over a decade, with problems of interest in the aviation industry like air traffic control, control of autonomous surveillance aircraft, logistics planning and scheduling, and equipment diagnostic and repair.[265] Our industry can transfer learnings from aviation, in a range of real-time decision making processes like performance monitoring of trading platforms and IT networks for anomaly detection.

It is expected that all future capabilities in an organisation will be built on AI, and most CEOs and decision makers will have their AI assistant listening in on meetings. Predicting outcomes is a useful prospect to have using AI. It is essential, however, that as the board uses AI more often they do not delegate its duties to such AI assistants.

Meetings

Board meetings can be transcribed and action points summarised. Such functionality is not solely the dominion of companies with deep pockets. FireFlies.ai is a free service that provides meeting recordings and transcription of meetings. Otter.ai transcribes meetings in real time, identifies speakers, and has a search functionality of meetings going back 90 days. Microsoft, IBM Watson also offers text-to-speech services, among which include automatic audio transcription from seven languages, rapidly identifying and transcribing what is being discussed, even from a lower quality audio, and presenting the words and phrases with confidence scores.

There are AI tools that make meetings more effective by taking meeting notes and issuing action points, following a meeting. Testfire Labs, a Canadian company, built Hendrix, which has the goal to make meetings more effective. Microsoft and Cisco have built similar tools, which also enable calendar and e-mail integration. Salesforce uses an AI tool designed to do forecasting and modelling. Salesforce CEO Marc Benioff said, "I will literally turn to Einstein in the meeting and say, 'OK, Einstein, you've heard all of this, now what do you think?' And Einstein will give me the over and under on the quarter and show me where we're strong and where we're weak, and sometimes it will point out a specific executive, which it has done in the last three quarters, and said that this executive is somebody who needs specific attention during the quarter."[266]

AI for Scenario Analysis (Modelling)

Scenario simulation is one of the most powerful tools to use when reaching a decision. Agent-based modelling (ABM) and multi-agent modelling are two methods that have been successfully used in studying financial and macroeconomic stability, as well as helping in assessing macro-prudential and monetary policies.[267] At the enterprise level, applied research has shown that the design of a data model by business executives is an efficient way of visualising complex problems through the transaction of verbal data into visual.[268] Advanced forms of ABM are being applied in portfolio management, workforce management, IT networks and distributed computing, consumer behaviour, and organisational behaviour.

AI LITERATE BOARD

Board directors have many fiduciary duties, among which are those of loyalty and care; they have the duty to act in the best interests of the corporation and its shareholders. They also have the duty of oversight, which requires that they put systems in place for effective reporting and monitoring to allow them to identify potential risks to the organisation. In the earlier section on Oversight of AI, I showcased in detail various components that are required for an improved oversight function. The duty of oversight needs to include their ability to nudge the CEOs to move the organisation forward at a fast enough pace to keep it ahead of the competition. The UK's Financial Reporting Board identifies the ultimate responsibility of the board: "Every company should be headed by an effective board, which is collectively responsible for the long-term success of the company."

Board composition needs to be aligned to the organisation's business aspirations and vision. A study found that high-performing S&P 500 companies were more likely (31 percent) to have at least one digitally literate board director than other companies (17 percent). Less than 10 percent of S&P 500 companies had a technology subcommittee. Compliance and digital risk management are currently just a check-the-box procedure, not a specific strategy. In the US, the Cybersecurity Disclosure Act requires publicly listed companies to explicitly qualify in their annual reports whether or not they have a cyber-security expert on their corporate board. In case they do not have such expert, they are required to explain why they do not need one. It is not long before we will see similar provisions for AI experts.

Having a digitally literate board is good business and the "new financial performance differentiator," according to an MIT Sloan Management report issued in 2019.[269] The research team found that companies with digitally savvy boards had:

- 38 percent higher revenue growth
- 34 percent higher ROA
- 34 percent higher market cap growth

While these numbers translate into a more profitable organisation, they also mean an organisation that is likely to offer security of jobs for their staff. But there's another important detail to bear in mind: These are annual numbers–over time they compound. Just compare and contrast the financial vitality for the next five years of an organisation with an annual 38 percent revenue growth compared with that of an organisation that does not implement AI. Sooner rather than later, the competitor gap will become insurmountable for those organisations whose boards have not given a strong mandate for AI adoption for business growth and profitability. AI strategy is good governance. Without it, boards leave their organisations with a business existential risk.

AI DIRECTORS ON BOARDS

The topic of AI directors has left many board directors slightly bewildered as to why they should accept such "odd" members to their board. What do I mean by AI directors? Here's an example: VITAL is an algorithm that was "appointed" on the investment committee of a VC firm based out of Hong Kong that specialises in biotech investments. VITAL has been operational since 2012/2013. It demonstrates that an algorithm can augment the quality of the decisions a board makes. Based on the information I have on VITAL, it is not a board director in the traditional sense, however it has a vote in the investment decisions made by the board.

An article about VITAL incited people's imaginations. It also extrapolated the discussion into the philosophical realm of organisations solely run by algorithms. That's why the IEEE AI design principles need to be embedded in how AI is deployed (see more details in Chapter 5: Corporate Governance and AI Adoption). However, I think that the current generation of directors will push back on the adoption of AI Directors but welcome the insights generated by an algorithm. Adding an AI director is a big jump–too different from how we have been doing things.

Moreover, there are no regulations in place to adopt AI directors on boards. Directors have many responsibilities. To put those responsibilities on to AI directors may mean entering the space of personhood, accountability of electronic persons, and legal rights for electronic persons (a true minefield). Legal rights for electronic persons is not something that I advocate, especially in financial services, because it creates the premise for an opaque "shell company,"[270] which is incongruous with the spirit of transparency and accountability the financial services regulators are promoting.

I think that there are sectors like biotech where the volume and velocity of data require AI tools to analyse the data. However, I am yet to see how algorithms would be able to make good decisions in private markets, where the data is usually in short supply and inaccurate. Actually, in March 2019 I recorded a podcast with William Tunstall-Pedoe, a serial AI investor and the co-creator of Amazon's Alexa, and we talked about this exact topic. He, like me, does not see how an algorithm can actually add intrinsic and reliable value in private capital markets.

However, it is likely that board will soon be making decisions based on real-time data. The volume and velocity of data, in addition to the need to extract insights and interpret data in real time, will require AI capabilities. In fact, my research indicates that JP Morgan's board of directors and executives are already using real-time data insights during their board meetings and use specific algorithms trained to provide insight (prescriptive, predictive but without autonomy of execution). Real-time financial projections are the core of any innovation of finance function with AI. Any successful business will need to consider investing in access to real-time information for real-time decision making. However, for this to happen, businesses need to meet one of the preconditions for AI: adequate computing power which most organisations do not have.

I think boards will continue to exist on a scale ranging from (1) boards who do not get AI to (2) boards who use AI to augment the human director's decisions. I do not think we'll see autonomous organisations run exclusively by AI directors (algorithms) anytime soon. If ever! I do think that the next urgent direction for businesses would be to build the hardware infrastructure to support access to real-time data analysed with AI. Such an infrastructure would be ideally designed using next generation hardware.

Finally, it is worth noting that the Information Commissioner's Office (ICO) in the UK has already made specific calls for boards to sign off on AI/big-data analytics projects. It will not be long until the financial services regulators will ask for the same.

LEADERSHIP

There is no doubt that in transformative times, it is helpful to think that the right leadership will show the way. During such times, it is vital to be realistic about the skills needed for an effective leadership. Such realism is also good governance.

Cognitive Finance Group's research identified that as of the end of 2017, approximately 94 percent of the leadership in financial institutions listed on the London Stock Exchange had no connection with technology–either in their current or past roles–nor did they describe themselves as interested in technology or having a background in mathematics, computer science, or sciences. It is relevant to review the skills needed at the board level to steer the organisation through the current wave of technology. It is absolutely essential and highly recommended that boards include at least one director, either executive or non-executive, who is able to bridge AI technologies with specific sector knowledge and corporate governance. These people are very rare, but they need to be brought over.

Most boards are not aware that the lack of technical ability to differentiate AI tools from pseudo-AI tools undermines the procurement and implementation of AI systems. Ninety-three percent of CEOs interviewed by IDC in 2018 found that they were not satisfied with the outcome of the investments made in digital innovation, including AI. This is not surprising. About 80 percent of the AI projects that I saw in pilot and production in financial services are used in isolation (specific process) and not as a part of the business strategy or value chain.

Information architecture needs to be adjusted to AI implementation objectives. For instance, almost 80 percent of the processes that were automated with AI were not redesigned for machine-centric design but followed manual-centric design. It's like trying to ride a bicycle with square wheels.

Skills optimisation in boardrooms made the focus of a research study produced in April 2018 by Harvard Law School Forum on Corporate Governance and Financial regulations. The research presented a machine-learning approach to selecting the directors of publicly traded companies and identified that "the algorithm is saying exactly what institutional shareholders have been saying for a long time: that directors who are not old friends of management and come from different backgrounds are more likely to monitor the management." [271] Naturally, this is a biased approach and diversity is required to maximise shareholder value. I see some merit in an *ethically designed* recruitment tool which, for instance, would be able to address gender diversity on boards, or any other aspect of recruitment where humans' biases are deeply rooted.

Bridging the AI Knowledge Gap

Among the board directors whom I interviewed, I talked to a chairman of three financial boards (mid-size UK branches, retail banking). He dismissed the idea of having an AI expert on his boards. He believes that boards typically discuss matters that do not need AI expertise. "The Board is not discussing IT. So, what's the point of having an AI specialist on my boards?" he asked me, incredulously. He was convinced that AI applies only in customer on-boarding and fraud prevention.

This is not an isolated case. Currently, there is a knowledge gap that may affect the delivery of the board and C-Suite responsibilities. It also impacts the governing roles of the board and its accountability for propagation of an innovation culture and of long-term thinking. Most financial services' leadership is unprepared for digital transformation. The changes are happening rapidly. I found that the most effective leaders I spoke to were those who, despite no technology background, displayed a level of intellectual curiosity and desire to find out more about how this technology works, all coupled with a level of humility and ego-less approach. They were leaders who would openly say "teach me what I do not know."

There is a mindset gap that created a number of blindspots for leaders. MIT, SMR, and Cognizant surveyed more than 4,000 managers and leaders from 120 countries. The results revealed that a mere 9 percent strongly agreed that their "leaders had the proper skills to compete effectively in the new economy." This report identified the four main blind spots:[272]

1. *Strategic* – leaders missing out on business growth opportunities
2. *Cultural* – leading by example was given as the most effective way to effect cultural change
3. *Human Capital* – leaders who are not invested in the professional growth of their staff will lose the forward thinkers
4. *Personal* – leaders insulating themselves from a changing world

Chief AI Officer and Chief Values Officer

Many organisations have already created a Chief Digital Officer role. I invite board directors to think ahead and consider that appointing a Chief Artificial Intelligence Officer on their board is almost an immediate necessity. AI will become such an important part of every business that it will be quite normal for companies of all sizes to have an AI Evangelist or a Chief AI Officer (CAIO) on the board. The role of the Chief AI Officer (CAIO) is taking shape in a key role in a number of financial institutions. TD Bank Group, Canada, has hired a CAIO and said that such an executive and member of the C-Suite is working across the enterprise with the mandate to set strategies, recruit AI experts, and cover all the bases to applyAI to improve business. In 2018 JP Morgan Chase hired a top AI expert who joined their corporate and investment bank as the head of AI Research. National Bank of Canada, the largest bank in Quebec, has hired a Chief of AI. From several discussions, it has become transparent that not everyone thinks that such a role is needed, because they misguidedly think that AI is an IT issue.

Unfortunately, the human talent pool for this role is scarce. Because a CAIO is the bridge between the business and data scientists, this person is the translator to the board of what AI means in business terms. This is leadership role, hence the board position for the business and for the technology teams. So, what is the professional profile of this person? He or she must be:

- a business growth expert, because there must be a clear alignment that the technology follows the business
- a risk management expert, because of the expectations to understand and work with the compliance department
- a business strategy execution expert, with proven execution experience
- knowledgeable of AI advancement and current AI techniques, able to work with engineers and data scientists
- deep sector knowledge (insurance, asset management, retail)
- excellent people skills

The CAIO is your AI voice of reason at the board level. It will take time for you to locate and convince your future CAIO to join your board as an executive. AI in business is complex. The board directors I interviewed shared a core and misplaced concern, whereby they equated the black box algorithms to AI. It takes some elevated knowledge of AI to rebut this misconception. Therefore, education is key, because without suitable education, the board risks underusing AI out of fear of overuse or misuse. The board is not able to give a strong mandate to their executive for business transformation with AI. That's how they miss out on business growth opportunities. This means that AI will be adopted as a piecemeal approach, with limited impact and high costs. This will drive chaos and uncertainty, which will burn people out and erode their trust in the board.

In addressing pressing ethical concerns, Kay Firth-Butterfield recommends the creation of a Chief Values Officer role, a term that she coined. This role "would be responsible for educating employees, preparing company policy, and overseeing the development of products that will directly affect employees' and customers' agency, identity, and well-being."

BOARDS TYPOLOGY AND DIGITAL TRANSFORMATION

In order to identify the patterns that keep some boards away from progress and development of their business for a sustainable revenues model in the future, it is useful to categorise boards as either "passive" or "active."

Table 6.3 Active Boards and Passive Boards Main Characteristics

ACTIVE BOARD	PASSIVE BOARDS
• Tests for risk, asks tough questions, pushes for honest conversations • Takes calculated risks • Uses data, requires quality data and insightful analytics tools to help their decision making • Works politics across the entire organisation • Makes data-driven recommendations for projects • Works with AI strategy experts specialised in their sector • Brings data insights into digital risk management	• Lacks and avoids measurable objectives in AI adoption • Hides behind a myriad of concerns which paralyses them • Relies on a library of rigid templates: standards, textbooks, processes • Shies away from conflict, politics, and truthful conversations • Standards and consistency determine how they decide • Drowns in endless reporting but does not generate ideas or direction for the executive • Has a backwards-looking approach limited to a present-only view • Avoids bringing in AI experts and directs with dissimilar background

Another useful framework[273] that enables a more granular analysis of the boards' behaviour is eloquently summarised in *Governance in the Digital Age* by Brian Stafford and Dottie Schindlinger, a book that I highly recommend for further reading on modern corporate governance.

Table 6.4 Governance in the Digital Age: Board Behavioural Profiles

	FOUNDATION BOARD	**STRUCTURAL BOARD**	**CATALYST BOARD**	**FUTURISTIC BOARD**
Description and focus	Meeting basic requirements and driving growth. Focus: **growth**	Board serves as a watchdog. Focus: **governance & oversight**	Board drives results in a "turnaround" capacity. Focus: **legacy**	Board is a strategic resource for long-term performance. Focus: **legacy**
Common traits	Board is small (< 5), with few independent directors. Informal processes.	Board grows in size. More independent directors. Process formalized. Governance staffing.	Directors drive change. Separate CEO and board chair. Focus on outcomes drives adoption of new processes.	Independent directors. Succession and board resilience are top of mind. Innovative board processes.
Board meetings and processes	Efficient, but informal. Few meetings and committees. Minimal documentation. No board support staff.	Meetings focus on reporting and discussion of results. Extensive board materials. Processes are better defined.	Focus on discussion of adaptation, positive change. Increased board education and committee engagement.	Directors drive the agenda. Meetings are strategic discussions. Directors create broad peer networks.

Table 6.4 Governance in the Digital Age: Board Behavioural Profiles (continued)

	FOUNDATION BOARD	STRUCTURAL BOARD	CATALYST BOARD	FUTURISTIC BOARD
Tech approach and tools to use	Tech tools don't play a promient role in the board's work - email and basic online repositories for documents might be in use.	The board begins to use online tools to facilitate meetings and automate compliance requirements, such as online resolultions, and board evaluations.	Catalyst Boards use online tools to facilitate meetings, track progress, and increase directors collaborations - such as agenda building, board peer evaluations, secure messaging, and online board education.	In addition to online meeting managemen, compliance and collaboration tools, Futuristic Boards want real-time content: dashboards and scorecards on areas of compliance, risk, and board effectiveness.
Outcomes	Board takes its lead from management for areas of untapped potential.	Board understands its role is to ask better questions; possible answers are still supplied by management.	Board acknowledges its role in company transformation, and becomes hungry for better data, analyses, insights and education.	Board understands its role as a strategic asset for long-term value. Directors adopt a "coaching role" with the CEO.
Key drivers to change profile	As directors gain knowledge and experience in governance, they see a neeed to exert more direct ovesight and develop more formal governance practices.	Directors feel they have at least a basic handle on compliance, and seek ways to work more collaboratively.	Directors foster strategic relationships with executive team; directors seek better insights, and begin to think about legacy.	Directors understand their strategic role, and can self-identify when it's time for them to step aside and make room for fresh perspectives.

EXAMPLES OF QUESTIONS FOR BOARDS TO ASK

	TOPIC	POSSIBLE QUESTIONS
1	Board's AI strategy	1. AI literacy on board? 2. Board efficiency with AI? 3. What is our strategy for continuous learning?
1.1.	AI Literacy on Board	1.How do we bridge the knowledge gap? 2. Do we have an AI expert to regularly brief us on AI developments relevant to our business/sector?
1.2.	Board Efficiency with AI	1. What AI tools can we use for helping us in our work? 2. Is our board secretary up to speed with these developments?
1.3.	Board Composition	1. Are there a minimum of two board directors with technology experience on our board? If not, how can we recruit them? 2. What is our succession strategy?
3	Business Reinvention	1. Has the executive delivered a list of new revenue streams to encapsulate the use of emerging technologies? 2. What are plans to set up our own R&D centre?
4	AI Oversight Review	1. Do we have a Technology Advisory Committee? 2. What is our process to monitor emerging AI and cyber risks?
5	AI Strategy	1. What is our AI strategy for the next 10 years? 2. Are we aware of the emerging AI so we can incorporate it into our AI strategy?
6	AI Adviser	1. Who is our business AI advisor? Do they have any conflict of interest in advising our organisation? 2. Does the Service Level Agreement (SLA) with them identify the conflict of interest?

KEY POINTS TO REMEMBER

- IT has grown from a mere operational infrastructure to be the core of business risk as well as strategy deployment and design. It is good corporate governance to regard IT, and any digital and AI-related developments, including AI ethics, as a board issue.
- It is important that the board ground its disruptive-risk oversight with a solid understanding of the company's key strategic drivers and of the significant assumptions made by management that underpin the strategy.
- AI can vastly improve the productivity and efficiency of board meetings, enabling board members to focus more on business strategy.
- Companies should invest in an AI tool for board packs that can help with document analysis, reading comprehension, and summarising key points within thousands of pages.
- AI can enable access to real-time data and reports to help with decision making during board meetings.
- AI tools can also transcribe board meetings and summarise action points, even identifying speakers and including search functionality.
- Scenario simulation is one of the most powerful tools to use when reaching a decision. Agent-based modelling (ABM) and multi-agent modelling are two methods that have been successfully used in studying financial and macroeconomic stability, as well as helping in assessing macro-prudential and monetary policies.
- Having a digitally literate board is crucial and can lead to higher revenue growth, higher ROA, and higher market cap growth.
- AI strategy is good governance. Without it, boards leave their organisations with a business existential risk.
- It is absolutely essential and highly recommended that boards include at least one director, either executive or non-executive, who is able to bridge AI technologies with specific sector knowledge and corporate governance.
- Many leaders do not have the skills or AI knowledge necessary for digital transformation.
- Board directors should consider appointing a Chief Artificial Intelligence Officer on their board. Because a CAIO is the bridge between the business and data scientists, this person is the translator to the board of what AI means in business terms.

CHAPTER SEVEN

Strategic Adoption of AI in Business

The financial services industry has entered a critical moment in its evolution where we must question where, how, and if current business models are ready for a profound paradigm shift. This is imperative, because digitalisation is reshaping how organisations maintain profitability and how they compete. The year 2019 marks a turning point in the understanding of how AI technologies impact organisations. AI must be understood properly to be used effectively. In his book *The Innovator's Dilemma,* the Harvard academic Clayton Christensen explained, "how good managers could wreck great companies by failing to embrace disruptive technologies. The efficiency with which they took decisions and invested resources to maximise existing opportunities only deterred them from exploring emerging possibilities.Their very success accelerated failure." Strategically adopting AI is a necessity for any competitive business.There is an indisputable body of research that reveals the benefits of adopting AI while there is still inertia in organisations:[274]

- Organisations that have already begun their AI journey are doing 5 percent better on factors like productivity, performance, and business outcomes than those that have not.
- Organisations investing in developing underlying values, ethics, and processes for their use of AI are outperforming those that are not by 9 percent.
- Fifty-one percent of UK leaders admit their organisation does not have an AI strategy in place.
- Just 18 percent of UK workers say they are actively learning new skills to help them keep up with future changes to their work caused by AI.

And in this chapter we'll look at key benefits like scalability, personalisation, speed, and more, what they mean, and how they are being put to use in financial services. Throughout this book, I encourage boards and decision makers to regard technology as an enabler to execute their business strategy vision rather than an accessory *du jour* driven by a meaningless "let's do something about AI" approach to innovation. Delivering a business strategy has many moving parts, but there is one that should never change: *Technology must follow the business strategy*. This poses questions in two important areas:

1. Timing: How do we know when it's the right time to invest substantial resources in an emerging technology? Are we buying into an extinguishing technology or are we getting in too early? Do we need to wait or get in now full on?
2. Knowledge: What do we need to know in order to ensure that our business vision makes full use of the emerging technology so that we remain competitive? How do we ensure that we know what to buy, and our staff is able to use it to its full capacity?

This chapter aims to answer these questions and take a closer look at ways to reframe the thinking about strategically adopting AI, based on the following themes:

1. Technology follows business strategy.
2. Organisations need to first focus on process and people, not technology.
3. How unprepared are the majority of leaders and boards for digital transformation?
4. AI is a business strategy issue, not just an IT issue–need a blueprint for AI strategy.
5. Innovation should ultimately mean an AI-first company.

We'll explore where to start with AI, covering essential points from educating your staff to reimagining your entire business including tackling internal resistance to change and choosing the right AI adviser. I'll explain the five main components of data strategy as well as the critical touchpoints for risk management. We'll also examine how AI changes your workforce strategy, including how it can be used properly and responsibly by human resources, the creation of new positions like a Chief Digital Officer, the reconfiguration of jobs, and building trust both within and outside of your company. Finally, when implementing AI, research and development innovation labs can be instrumental for testing your AI tools until they work as intended.

ADAPT OR PERISH

There have been many changes in financial services in the past four decades. "You never stop learning in this industry," Nigel Thomas, one of the most respected fund mangers, notes in his widely circulated Thomas Report of AXA Investment Management in London. Mr Thomas referred to a myriad of economic events that shaped his investment philosophy while he also concluded, "The new rules of data capitalism are still being written. Where data capitalism is embraced in data rich markets, traditional companies are challenged and have to adapt or become obsolete.[275] Let's take a look at what Mr. Thomas says in his last Thomas report, where he highlights why to avoid buying companies that do not adapt:

> "It is one of the functions of the investment manager to [...] avoid the 'dinosaurs.' It should come as no surprise that Woolworths and British Home Stores no longer exist. However, looking at the list of consumer companies that are, or have been, in administration or Company Voluntary Arrangements (CVAs) is perhaps an indication that the Amazon effect is taking its toll–Homebase, Mothercare, House of Fraser, Maplin, New Look, Carpetright, Toys R US, Poundworld and, more recently, Coast. This influence of online platforms is not restricted to retail, and extends to other industries such as property, automobiles, travel and hotel accommodation. Within [our fund] we own Rightmove, Autotrader and also Worldpay, who facilitate transactions in the digital world. It took Airbnb three years to enter 89 countries, and took the Hilton Group 72 years to enter 69 countries. In 2007, CEO of Microsoft, Steve Ballmer, pronounced that 'there is no chance the iPhone is going to get significant market share.' By neglecting the steady and continuous advancement in computing power and mobile technology, Microsoft missed the smartphone phenomenon."

If investors lose faith in an organisation that is not innovating fast enough, they'll sell that stock. The knock-on effect on the share price impacts retail investors and pension funds, and ultimately it reverberates negatively in society in one form or another. This is why AI strategy adoption is good governance. Business transformation with AI goes deep into the fabric of a business with ramifications in human resources, infrastructure, and management style.

As early as 2011 it had become clear to me that the available technology wasn't good enough to cope with the deluge of data, and the rudimentary infrastructure typical in our industry wouldn't be sufficient to sustain such ample volumes of data. Such infrastructure was and still is expensive to maintain. It is unfit for the purpose of supporting the technological advancements expected from a modern business model that assumes that technology permeates every layer of a business offering. It's hard to predict the future, but I think on this occasion Gartner might have gotten it right: By 2030, the enterprises that commit dedicated organisational resources to ensuring that strategy is successfully executed are 80 percent more likely to be industry leaders. Some 80 percent of the "work" that represents the bulk of today's project management discipline, practices, and activities will be eliminated by partnerships between humans and AI.

AI DEFINITION IN FINANCIAL SERVICES

For business purposes, it might be useful to articulate what AI is in the financial services market. In its seminal report "The New Physics of Financial Services," the World Economic Forum provides a useful framework to define AI: "a suite of technologies, enabled by adaptive predictive power and exhibiting some degree of autonomous learning, that have made dramatic advances in our ability to use machines to automate and enhance:

1. Pattern detection–Recognise (ir)regularities in data
2. Foresight–Determine the probability of future events
3. Customisation–Generate rules from specific profiles and apply general data to optimise outcomes
4. Decision-making–Generate rules from general data and apply specific profiles against those rules
5. Interaction–Communicate with humans through digital or analogue mediums"

This framework can help us begin to see how AI can benefit certain business sectors and where it might be helpful to implement AI tools.

WHERE TO START WITH AI ?

Where do we start? I get asked this question many times, especially by overwhelmed executives who have already started with AI but are yet to reap the benefits. The 10-step framework below can help you figure out how to get started, some of the key initiatives to focus on, and how to deliver a strategic approach:

1. *Educate your staff.* The Board, CxO, and your entire staff (and yes, that includes your receptionists, too) ought to know what AI technologies are and how they can be used to reinvent your business models. Chapter 2: The Taxonomy of AI and Chapter 3: The prerequisites for AI, respectively, form the foundational ground for a thorough

understanding of AI technologies, the infrastructure they need, and their various applications. For instance, the Oxford Artificial Intelligence online course[276] and the MIT AI in business courses are excellent launch pads for educating your people in large groups without too much administrative and cost burden. *Core question:* Does everyone in my organisation know what AI is and do they have an understanding of the possible applications in our organisation?

2. *Reimagine your business.* At this stage, you need to bring everyone in and brainstorm where they would like to see their organisation in the future. Everyone means everyone. Yes, everyone includes your receptionists, too. This is all about horizontally engaging and opening up channels of communication with your staff. Do you think this is a waste of time? It isn't. You are reimagining your business, and you need to take your people with you. Breaking down silos determined by departments and internal politics is your main objective, in addition to aligning your people to your vision. You need to articulate business goals. Some examples would be (1) to have a presence in 60 countries in 10 years or (2) gain 20 million clients in five years in India. Do not focus on costs at this stage; focus on your vision for business growth. *Core question:* Where do we want to be in 10 and 20 years time?

3. *Choose your AI adviser.* This is possibly one of your most important decisions to make. Your adviser needs to have proven core competencies in applied AI and sector-specific knowledge. He or she does not need to be one of the big consultancies–please do not fall into this trap. Your adviser needs to have a proven track record of specialising in *your* sector, knowing the AI start-up community pretty well, and be up to speed with where the AI research community is heading, so you know you are buying the next generation technologies, not already outdated technologies. Your adviser should be able to bring in the right specialists to serve your specific needs: big-data experts, cloud experts, AI experts, and strategists. Just a fact worth noting: Large consultancies outsource most of the services they charge for. Their past experience is not necessarily an accurate predictor of their future success. Your Ai adviser will be able to work with you to build a budget for this transformation budget. *Core questions:* Is our AI adviser impartial? Does he or she have vested interests in selling us a specific product or services? Does he or she take a commission from any recommendations?

4. *Upgrade your board.* You need to immediately add at least one Non-Executive Director to your board. The job description needs to also include as requirements: (1) proven experience of applied AI, (2) sector-specific knowledge of either current

or former practitioners, (3) deep knowledge of corporate governance, and (4) no need of prior board experience. You need people who bring a fresh look at your business through technology lenses. Former or current consultants from large consultancies are not recommended due to potential conflicts of interest. *Core questions:* Do we have an AI literate discussion at the board level? If not, do we need to set up a Technology Advisory Board or perhaps hire another AI expert as a Non Executive Director? Where do we need to speed up succession plans to make room for an AI literate member? Are we clear at the board level that we are guiding the organisation into the next transformational stage?

5. *Design your AI strategy.* Remember that technology follows business strategy. At this stage, you have internal capabilities to engage with your AI adviser. You have at least one member of your board who will be able to support your vision at the board level. You have hired an AI adviser who will be able to translate your business goals into technology tools. (The remainder of this chapter goes into detail on the core considerations for an AI strategy.) *Core questions:* Are we clear that we are building a blueprint for business transformation? Where are the "politics" pain points and how do we remove them? Have we reassured our people that they know where their new roles will be as the organisation is shifting to a new business model? What technology do we need to take us where we need to be?
6. *Redesign your information architecture.* At this stage you need to be able to take an all-encompassing look at how your organisation is structured, or how the data flows within the organisation. You also need to be able to break down each business process to its component parts. Finally, you need to establish if individual business processes can be automated or not, and once automated how they will work and impact the overall organisation. *Core questions:* Do we understand how each business process is currently intertwined? Do we know which processes are designed for manual delivery and how we redesign them for automation? What new processes do we need to introduce (e.g automated voice recognition, document summarisation, etc.)?
7. *Identify your first AI project.* This needs to be part of your new AI strategy; it has to have meaning in the larger organisation's workflow. It has to be small enough to be completed without surprises, but meaningful enough to demonstrate how it can make a difference (e.g. cost reduction or increased accuracy). *Core questions:* Where do we need to start? What's the simplest process that we can automate with AI, and what technologies do we use?

8. *Select your AI provider.* Your AI adviser can advise what makes more sense for you–to buy or to build AI. Check your vendor selections criteria, which should also include technology assessment criteria; system scalability, explainability, accountability, data and AI model assessment; privacy capabilities appraisal; regulatory benchmarks matching; operational processes' impact on business; scaling up; and change management. *Core questions:* Is this actually AI? Can it be scaled? What do we need to scale it?
9. *Test your AI.* At this stage you need to appreciate that building a pilot and then putting it in production is a laborious and time-intensive exercise. The systems need to be back-tested, refined, and verified on various assumptions and data. This requires time and management's patience, and if done correctly it will pay off in the end. It might take anywhere from eight months to 28 months to see a model live. Core question: Have we ensured that verification is solid and the model is ethically designed?
10. *Deploy and scale your AI.* Hard work pays off. You are now ready to deploy your first model live. You need to test a range of launch pad assumptions, which are discussed later in the chapter: (1) technical; (2) brand; (3) governance; (4) organisational. *Core questions:* How can we turn off this system in case of emergency? Who is in charge? And how can we monitor if it operates within the normal parameters? How quickly can we scale it? Do we have the right AI infrastructure in place to scale it ?

WHAT TO DO WHEN THERE IS INTERNAL RESISTANCE TO CHANGE

One main reason why people are reluctant to change is that they do not know what will happen to their jobs. This uncertainty paralyses many and causes them to stay still in the hope that stillness will not upset the status quo.

In 2016 Martin Ford's book, *Rise of the Robots: Technology and the Threat of a Jobless Future,* became an overnight best seller–not because it brought some unique angle to the technology revolution but because it eloquently articulated the fear about a jobless future. Fear sells. The art of fear-mongering was supremely well delivered in this book. The media swiftly picked it up and magnified it.

At the beginning of 2017, the majority of the financial services community was deeply entangled in a web of fear. On this backdrop, it is not surprising that many executives felt that they shouldn't be seen talking about introducing AI in their organisations in fear of scaring off their staff. This fear still persists, albeit more subtly. It continues to drive people to consciously or subconsciously undermine AI adoption projects. And so the core question is: How do we remove the fear by empowering people to see what's in it for them? One successful way to achieve this is by adopting AI tools to solve the staff's typical time-consuming tasks. Consider adopting an AI tool able to coordinate diaries, summarise meeting notes, distribute action points and follow up with each member of the team, take notes, generate action items, and track time spent on projects. Another way is through curated training disseminated across the organisation to teach people new skills so they can stay relevant in their jobs. As long as people see what's in it for them, they will move forward with AI adoption to solve other problems like customer engagement and redefining business models. Not all jobs will stay as they currently are: Some will be redefined and some will become obsolete, and new jobs will appear like AI ethicists or AI forensic analysts.

An executive told me that they will not adopt AI because "the majority of their staff will feel threatened by the job losses." Organisations adopt AI in order to be able to compete and remain profitable. When organisations are not able to compete and remain profitable, they go bankrupt and *everyone* loses their jobs. The choice seems pretty simple.

DATA-DRIVEN VS. AI-DRIVEN BUSINESS

There is a deep confusion between these two terms, and it's important to know the difference when you are building your AI strategy:

- Data-driven business describes a focus on data collection, storage, and manipulation, with little focus on processing data for insights, decision making, and business growth.
- AI-driven business describes a focus on data processing and insights extraction to support decision making for business growth. In an AI-driven business, data processors are both humans and algorithms. It is important to make the distinction between their respective abilities, and they should be used where their respective abilities deliver value. Humans' processing capabilities are limited, hence the *augmented intelligence* and *man and machine working together* narrative. Algorithms have superior capabilities to process data, draw insights, and make decisions. However, long-term decision making is usually better left to humans who only need the insights.

AI ADOPTION: KEY BENEFITS

AI in financial services is an important technological advent: For the first time in our industry we can provide personalised services at scale and in real time. This opens up new business growth opportunities and efficiencies. AI is a key enabling technology that provides a series of tangible benefits that have profound implications on the enterprise's business model, its clients, and indirectly the society.

In defining a new business model, AI will introduce new commercial success variables like the size and ownership of data. The speed of change will leave many organisations behind, and some of the laggards will not continue to exist.

On-demand insurance is an example of how a subscription payment model tailored to clients' needs has proven sustainable and gained traction. According to Kaggle, professionals in finance and insurance spend 50 percent of their time collecting or synthesising data, the largest rate across all other industries, leading even the government sector by 16 percent. This is not surprising. Finance and insurance have historically been data-intensive industries. The last decade's technological and digital advancements have increased the volume and velocity of data generated within and relevant to this sector. If you add to this an ever-increasing volume of regulations, and no adequate investments in intelligent software to process this increased volume of data, you start to understand why the industry needed to compensate by hiring more people in compliance and back office. AI impact on jobs will be the greatest where the largest amount of time is spent on collecting data, analysing it, and synthesising it. This is why the number of jobs lost in financial services will be substantial. This will lead to a trimming down of the workforce, improved efficiencies, and will also reshuffle the market players.

"Leading with Next-Generation Key Performance Indicators" is an original thought leadership research report produced by MIT Sloan Management review. Its authors argue that key performance indicators (KPIs) play a critical role in improving corporate leadership. They demonstrate how AI can impact and redefine strategic metrics and empower leaders with the knowledge of which KPIs to optimise. They conclude that AI might be better suited to create business strategy.[277] This remains to be seen.

SELECTING YOUR AI ADVISER

Institutions prefer to choose large, well-established names when they decide to embark on AI adoption. "I cannot be held responsible, if I choose a well-known, established name," one of the executives told me, shrugging his shoulders. "I don't want to lose my reputation. If we appoint a large consultancy, it is simple: They get it wrong, they take full responsibility." A few weeks after this conversation, I came across an article that covered in detail how Hertz, the car rental company, appointed Accenture to upgrade their website and to embed a range of analytics. Not a complicated job, for any small or midsize company. And yet, Accenture didn't deliver anything apart from a catalogue of errors and delays, and Hertz is now claiming $23 million in compensations and repayments of the fees they have already paid.[278] This is a compelling example for others in similar situations that never reached the media. The lesson is the same: Your AI adviser must be independent and with proven experience and knowledge in the design, selection, and delivery of AI. Unfortunately, organisations and their boards prefer to appoint a household name as their AI adviser. It is their default option. According to IDC, in 2017 93 percent of three thousand CEOs interviewed were not happy with the outcome of digital transformational programmes they invested in and according to McKinsey 75 percent of digital transformation projects fail. Why is this the case?

However, some large consultancies take a commission from the sale of the technology they recommend. "They recommend what is available today and yesterday. They don't talk about tomorrow," a director tells me with an irritated and exhausted voice. "We have had enough PowerPoints and juniors sent through. The world is moving fast and they are not helping us to move faster." So, they have a vested interest in recommending a certain product. If their advisory clients purchase the AI product, the consultancy will take anywhere between 10 percent and 30 percent of the value of the sale. It is no wonder that with some products you see the same error repeated in organisations across the same sector.

AI STRATEGY

AI strategy means reimagining your business with AI, so your business strategy becomes your AI strategy. The absence of a coherent AI strategy is a business risk. Why is it important for boards to put AI technologies at the heart of business strategy for the next decade? Technology must follow the business strategy and not the other way around. AI adoption is more than IT transformation–it is about reimagining how a business generates new revenue streams and reaches increased productivity.

Unless and until boards and the C-suite understand what AI technologies can do for their organisation's growth, they will not be able to weave AI into their business strategy in order to unlock AI's full value. AI technologies are becoming a defining factor for retaining clients and protecting market share. For instance, Marcus, the GS retail brand and a new entrant in retail, operates autonomously without human intervention and has registered almost instant success in the UK (retail banking) and the US (consumer lending). Cognitive Finance Group's analysis has identified that the absence of a coherent AI strategy is an existential risk to businesses operating in financial services.

What Is ROI on AI investments?

This might not be the right question, and yet most executives believe that this is the right place to start. Business transformation with AI requires patience and capital. It takes time. A valuable correlation between technology implementation and organisational change was explored by a group of MIT professors who concluded that "each dollar of computer capital was often the catalyst for more than 10 dollars of complementary investments in 'organisational capital' or investments in training, hiring, and business process design."[279]

Professors Brynjolfsson and Hitt, of MIT, analysed the effect of computerisation on productivity and output growth using data from 527 large US firms over 1987–1994 and concluded that "companies with the largest IT investments typically made the biggest organisational changes" and there would be five to seven years before they would be able to see the full benefits like business performance benefits.[280]

Reimagining Your Business Model

The 10-step approach to AI adoption discussed at the beginning of this chapter highlights r*eimagining your business* as step 2. The core teaching here is that organisations need to be bold about what they want to achieve. They can be bold only if they understand what emerging technologies can help them with (step 1).

There are a number of successful cases where the current business models have been pivoted away from traditional business models. For example, Facebook shifting into payments over WhatsApp seems like a move away from their traditional social media focus, yet it is an organic extension of their social media-centric delivery. Tesla, the automaker, announced in 2019 that they are moving into autonomous taxi services, undercutting Uber and Lyft prices in the $2 to $3 range per mile.[281] There are around 400,000 Tesla cars on the road, collecting billions of miles of driving data needed to train its fleet. In contrast, other auto makers struggle to collect the same volume of data to train their cars in autonomy. Tesla has secured first mover's advantage in autonomous driving.

Table 7.1 Main Benefits of AI Adoption

BENEFITS	WHAT DOES IT MEAN?	EXAMPLE IN FINANCIAL SERVICES
Scalability	increase reach to new markets and clients	digital assistant and automated client interaction
Personalisation	tailored customer service to what the client needs	customer retention and new business development
Efficacy	produce the intended outcome as planned with limited margin of error	optimise trade execution
Cost reduction	delivers the same outcome with no impact on service quality	signals for uncorrected returns
Speed	finishes tasks rapidly	fraud prevention
Accuracy	completes task with reduced margin of error	assess data accuracy
Efficiency	achieves maximum productivity with minimum effort	portfolio optimisation synthesising data
Productivity	improves productivity in jurisdiction where human capital is scarce (e.g. the Isle of Jersey)	back office intelligent automation risk management

In our industry, Goldman Sachs is possibly the best example of how to reimagine a business model. Since 2009 they have been paving the way to a leading position in autonomous banking. In 2009, when the whole world was picking itself up from the financial crisis rubble, GS made a strategic decision to move into consumer banking, a move which would give GS, the largest US investment bank, a more stable source of funding and revenue model. This was a move away from their institutional investment-banking-focused business model:[282]

- In 2012, they became a minority investor in the Australian Pepper Home Loans, acquiring GE Capital's Irish mortgages.[283] This was the first-ever transaction of this type GS committed to.
- In 2015, GS concluded the purchase of "GECB's online deposit platform and assumed GECB's approximately $8 billion in online deposit accounts and $8 billion in brokered certificates of deposit for an expected total of approximately $16 billion of deposits at closing."[284]
- In 2016, GS launched Marcus, the consumer-lending platform.
- In 2017, Goldman Sachs Bank USA was rebranded as Marcus.
- In 2018 Marcus launched its savings account in the UK, offering a competitive 1.5 percent interest rate well above the market and signing up more than 50,000 clients in two weeks.
- What is less talked about is that the entire Marcus proposition is designed to be autonomous, meaning to become the first Western European bank running without human intervention.

Business Transformation Means New Business Architecture

Boards need to understand the business architecture when discussing AI adoption because "business architecture is to an organisation what a blueprint is to a building."[285] Business architecture "represents holistic, multidimensional business views of: capabilities, end-to-end value delivery, information, and organisational structure; and the relationships among these business views and strategies, products, policies, initiatives, and stakeholders."[286]

CEOs must personally drive AI projects. They have to work closely aligned with CIOs, CROs, and CFOs. Reimagining business models impacts the functioning of any organisation in at least four areas: business processes, products, people, and insights. This is equally applicable to the front office as well as the back office functions. "AI allows humans to reach out beyond their intellectual limits or simply avoid mistakes"[287] and we have seen dramatic improvement in a wide range of functions. I would argue that, ultimately, any digital transformation is in fact a business turnaround. Stuart Slatter and David Lovett in their book "Corporate Turnaround" provide the a timeless framework to execute a successful turnaround. The book lists the steps to execute execute a successful turnaround (1) Crisis stabilisation; (2) Leadership; (3) Stakeholder support; (4) Strategic Focus; (5) Organisational change; (6) Critical process improvements; (7) Financial restructuring

In digital transformations, the speed of change requires the ability to adopt and adapt in order to fine-tune the new business architecture. Some call it agile transformation. Gartner provides a useful framework to evaluate an organisation's ability is to remain agile.
I would add that if organisations go through these steps rigorously and frequently, and if they pay close attention, they will also be able to find new ways to imagine their business models:

1. Continuous flow of business value
2. Frequent customer interaction
3. Iterative and adaptive approach
4. Creativity and innovation
5. Shared responsibility for results
6. Situational specific strategies

In a sea of changes and sometimes lack of visibility, innovation strategy management is a core function in the delivery of complex projects. GainX innovation strategy management platform leverages powerful artificial intelligence and behavioural analytics, allowing banks to bring all innovation resources. With the knowledge gained from the GainX platform, financial services companies can smoothly and repeatedly move from idea to market impact and quantifiably measure return on innovation investments (ROII). GainX and Microsoft will jointly work with leading financial services companies to help them envision and implement the integrated solutions.[288]

Data Strategy

In order to design their data strategy, organisations might need to start envisioning the value that they can strategically derive from the data they already have access to:

Table 7.2 Data Strategy Objectives

INTERNAL OBJECTIVES	EXTERNAL OBJECTIVES
1.business insights	1. monetisation
2. decision making	2. client engagement
3. operations	3. public presence

An example of monetising data as a new source of revenue was Goldman Sachs' announcement in May 2018 that they were getting into the business of monetising their own alternative data. Driven by the core of their securities division to identify the alternative data valuable in both investment and risk management, the bank looked at ways to monetise its internal data, while being conscious of client concerns around confidentiality, and so ensuring it was thoroughly anonymised.

Another example of monetising data may be for a consumer bank division to build an identity vault for their private clients. These vaults would enable clients to passport their identity digitally.

Defining a data strategy is one of the foundational pieces of the AI strategy. Data strategy has five main components and all converge to address "planning for how to improve all of the ways you acquire, store, manage, share and use data" and to "comprehensively support data management across an organization":[289]

1. Identify data and understand its meaning, regardless of structure, origin, or location. This includes management of associated metadata (definition, origin, location, domain values, file's author, etc.).
2. Store data in a structure and location that supports easy, shared access and processing. It is essential to efficiently manage the storage required.
3. Provide data to package and share the data according to the business units' needs. In organisations that regard data as a corporate asset and a value generator, data provisioning is treated as a standard business process.
4. Process data that is generated internally and externally to transform it from a "raw ingredient into a finished good." A significant role is played by unstructured data. Companies like AntWorks made it their mission to address a range of challenges with unstructured data and their approach has proven superior with clear immediate wins in accuracy and speed.
5. Govern data focuses on (1) establishing, managing, and communicating information policies and mechanisms for effective data usage and consistent management, and (2) rules and details of the data which must be known and respected across the organisation. Data governance is a key element of data strategy and initially addresses specific tactical issues like accuracy, terminology, standards, and business rule definition, which will just broaden in scope as the organisation's data governance matures.

The volume and velocity of data will only grow. In fact, Gartner predicts a staggering 800 percent growth in the volume of data between 2018 and 2023. Data processing and, in particular, unstructured data processing requires a particular focus. It is estimated that about 80 percent of the data in an enterprise is unstructured. This includes webpages, legal documents, back office records, mobile content like text messaging in WhatsApp, or Bloomberg. Therefore, it is essential that unstructured data is processed and distilled into a meaningful structure for AI algorithms to work with it.

Business Considerations when Choosing Cloud Providers

We have seen that there are a handful of cloud providers and all are making the right investments to provide ever more sophisticated AI solutions to their clients. In November 2018, AWS announced their latest products. AWS is currently the world's biggest public cloud, generating $6.68 billion in revenues for Amazon.

Some of their customers have become very concerned now that AWS is providing products that compete with their own. While this is currently applicable to other tech providers, it is strategically wise to realise that AWS is making a solid push into asset management and other financial services with enticing cloud storage offerings. As Amazon expands into countless new areas, from groceries to healthcare, some companies will inevitably find themselves competing with Amazon. Therefore, it is important to choose independent cloud storage advisers who will enable you to choose the right cloud strategy to not only keep your costs optimal but also protect some of your knowhow. In 2019, at a risk management conference in London, an entire panel formed of CROs of leading financial services companies discussed their disaster recovery strategy. All of them voiced the same concern that cloud providers were not transparent enough to satisfy FCA regulators' requirements and also didn't show any signs of willingly complying with industry requirements.

Risk Management

Another core pillar of AI strategy is the risk management framework. There should be a number of touchpoints with the business's general risk management framework and specific IT strategy risk management. In order to expand the reach and depth of discussion, I recommend the 2018 CRO Forum's "Understanding and managing the IT risk landscape: A practitioner's guide." It contains a range of exampled controls and key risk indicators, as well as framework assessments.

As AI tools are being deployed and integrated within the business's information architecture, there is an ever more important call on the resilience of IT infrastructure. An ineffective IT strategy, dated infrastructure, failing IT, and a disjointed information architecture are the most significant risks facing the board of directors. The following three pieces of the strategy puzzle are critical:

1. *Strategy Definition*–assessment of the strategic risks internally and externally. This includes a high-level examination of the strategic intent, followed by a close assessment of the strategic risks (internal and external). This is usually determined by the definition

of the business's risk capacity and by setting the business risk appetite. It also defines the parameters for information architecture (IA) resilience and processing and its efficiency.

2. *Strategy Implementation*–the overall business strategy is streamed into specific business objectives across business units, followed by a close evaluation of the information architecture analysis of how and if it is suitability designed for machine automation delivery rather than human delivery. Such analysis must include the definition of each business unit/function risk appetite and key risk indicators. AI deployment needs standardisation. Otherwise, it will become another legacy issue. The AI interaction with other systems and applications will dictate dependency. This raises the operational risk.
3. *Strategy monitoring*–the implementation of AI strategy objectives and plans needs to be monitored throughout its delivery cycles. Ongoing evaluation of the progress against the objective, monitoring the key risk indicators, revision of the proposed mitigation actions, and finally escalation of any breaches of risk appetite.

Technology Business Management (TBM)

TBM is a useful framework "instituted by CIOs, CTOs, and other technology leaders, founded on transparency of costs, consumption and performance." The TBM taxonomy was reviewed by The Technology Business Management Council (TBMC) formed by a number of leading and experienced CIOs. It is based on the thorough and extensive research done by APPTIO, the technical adviser of the TBMC. The TBM taxonomy the first generally accepted approach for translating finance, IT, and business perspectives. The TBM council published a book titled *Technology Business Management: The Four Value Conversations CIOs Must Have With Their Business,* which should be essential reading for decision makers.

WORKFORCE STRATEGY CONSIDERATIONS

In addition to the other components of your AI strategy, you must also consider the ways that your workforce strategy may need to change when adopting AI. This may include the creation of new positions within the company, reconfiguring certain jobs, and training employees in new skillsets. Company processes–both internal and external–will also likely change.

Personnel

One of the most important conversations is to explain and demonstrate to current staff that an AI-driven organisation is all about new skills but the same people. Most jobs will be reconfigured. This thesis was demonstrated in a paper by Professor Erik Brynjolfsson and Daniel Rock, along with colleague Tom Mitchell of Carnegie Mellon University, who studied the potential for AI to remove jobs across various sectors. Interestingly, they found that jobs will be redefined across all levels of income.[290]

Redefining roles requires redefining the HR strategy, and in my work this is an integral part of the overall AI strategy. In Chapter 9, I discuss various use cases of AI adoption. Some have proven to increase efficiency; some are simply a blatant invasion of privacy. Financial services has been a systematically surveilled environment. From recording all landline calls to monitoring WhatsApp messages, one might be able to understand the need for this granular level of scrutiny and regulatory compliance. However, I noticed a troubling trend in how the heads of HR regard the use of AI. One board member, and global HR head, believes that using biometrics and computer vision, and sensors placed across office and meeting rooms, is actually a great tool to have a detailed insight on staff because when "you have over 100,000 staff worldwide, you need to be able to control them and know what is happening."

In response to my question about invasion of privacy, I received a rather uninformed response from this HR expert: "Data points are anonymised, we survey the room not the individual, and we draw insight from a group not an individual, even when we use sensors to monitor their body language and body temperature to identify stress and discomfort and correctly identify their mental state." No data points can be fully anonymised. Once you are able to cross reference data points, which you can in a surveyed office space, there is no privacy left. My research has revealed that in some offices, restrooms are also surveilled, but the data collected is accessible only to top-level executives. This case goes beyond privacy issues–it undermines basic human dignity. WeWork is proud to announce that they use AI to optimise the activities in their office spaces, which they rent. Many fintech companies in London have chosen their office space in a WeWork building. I stopped them, as soon as the reception asked to photograph my face as a non-negotiable prerequisite to access the building. This has to stop. Some argue that it might even be illegal. This level of surveillance is detrimental; it undermines people's well-being, mental health, trust, and privacy.

Boards would be well advised to ask hard questions about deployment of AI tools that allow surveillance and prediction of people's behaviour (like IBM's HR tool, which is able to predict with a high level of accuracy when people intend to leave the organisation). Such practices are not only an assault on people's basic right to privacy but also when misused leave companies wide open to reputational risks that are hard to manage. Let's not treat our staff as lab rats. Some might think that the expected way to deploy technology is to imagine how they can replace each worker with a machine. This is flawed thinking. A number of executive roles are emerging and the new HR strategy should reflect them accordingly:

- The AI Engineer Role
- The AI Data Manager Role
- The Business Leader and AI Translator Roles[291]
- The AI In-house Educator

A number of new managerial positions have already emerged. For instance, Bank Corp created the AI Innovation Leader role. Many other organisations have already created a Chief Digital Officer and Chief Artificial Intelligence Officer (CAIO) role. It makes strategic sense to consider hiring a head of data, a Chief Digital Officer (CDO). This role's main responsibility is to look after one of the most important assets in your business: data. Most of the boards might need to create this role in recognition of how their business model is transforming into a data-driven business model. It's difficult to find quality candidates for this role, because it ideally encompasses two main knowledge sets with a fine balance between the two: (1) business background with deep knowledge in the sector which you are operating (2) technical background, or proven track record of knowledge in the field of data science, machine learning, and increasingly more important, deep learning.

The job description of a CDO may look something like below,[292] and due to its important role, it should be an executive committee seat or even board. This is yet another key role in any organisation, alongside with CAIO and CIO. A CDO has a business data-centric focus that includes:

- High-level vision–brings innovation technology and business together and understands why technology follows business strategy
- Implementation of board strategy at every level and C-level vision data accuracy, security, and privacy
- Identifying business opportunities–how to increase revenues and reduce costs
- Being a data-driven culture leader
- Seeing data as commodity – recognising value in data, either existent or potential

A number of new job specifications have already emerged. A few examples include digital trainer, storyteller, conversational brand strategist, algorithm forensic analyst, ethics compliance manager, digital product manager, workplace technology manager, and cybersecurity manager. However, deep analytics skills will continue to be in high demand. The International Data Corporation (IDC) has already identified a demand for data management roles that is five times more than qualified candidates.

Processes

Redesigning an organisation to become an AI-driven organisation is a redefinition of every single process. The following is a useful framework adapted from Gartner, the research and advisory firm, which highlights the key processes (internal and external) that change but that also interact and transform each other in a continuum.

Table 7.3 Key Processes–Change, Interaction, Transformation

INTERNAL PROCESS	EXTERNAL OBJECTIVES
1. AI implementation roadmap	1.Competitors
2. Business aspirations and goals	2.Ecosystem
3. CIO role and influence	3.IT systems + legacy
4. Data strategy to include procurement, preparation, monetisation	4.Customers
5. Sourcing, procurement, and vendor management	5.Brand
6. IT Infrastructure and operations, including information architecture	6.Regulatory requirements
7. Enterprise architecture and technology innovation	
8. Security and risk management	
9. Applications – process specific use cases and how they interact with the evolving business model.	

Trust

It is perhaps unusual to bring trust into a discussion that is purely focused on business strategy with technology. Trust with technology will be a theme that we'll hear many times. There are a few types of trust that AI implementation will redefine or reinforce.

The Trust of Your Staff

Between a misguided media frenzy about how "robots" will take over people's jobs, a reported lack of trust in their leaders' capabilities to lead the organisation in the new technology era, and the invasive use of AI in surveilling them while at work, gaining the trust of your staff is pivotal in delivering the change and convincing people to embrace new ideas. Boards and decision makers must be mindful about how to approach it. Rachel Botsman provides a valuable point to reflect on: "How to trust new ideas? How people take trust leaps? People do not want something entirely new. They want the familiar done differently." Rachel Botsman is an expert in trust and the best-selling author of *Who Can You Trust?: How Technology Brought Us Together and Why It Might Drive Us Apart,* a book that I highly recommend.

The Trust of Your Clients

The financial services industry is based on a core fundamental value: trust. The 2008 financial crisis left the world and our clients with a bitter taste about how we conduct ourselves when we are expected to act with the utmost integrity. For the purpose of articulating this point, I will include a useful definition of integrity: doing the right thing when no one watches.[293] As we adopt AI in this industry and look to redefine how we operate, now is the right moment to redefine trust with our clients–how we design the product they buy, how we put them at the centre of our delivery, and how we use technology to shift their perception of us from being a necessity in their lives to being a source of value that is run on human values. There is another type of trust that we still need to gain: digital trust.[294] As client data becomes more valuable, boards need to evaluate ways to derive value or commercial profits from it while keeping it safe. "For every interaction where data is shared between a private individual and an organisation, there is an implicit zone of trust created between the parties."[295] I anticipate that as people become more aware of how their data is being used and how organisations derive commercial profits from it, they will expect data accountability from these organisations.

The Trust of Your Suppliers

In short, this trust stems from being seen doing the right thing to build and maintain a successful business. This trust is constantly re-evaluated and closely scrutinised. It is formed by the talent and teams that you are able to attract, the progressive board that you are able to coordinate to lead the change, and how you derive value for your organisation.

The Trust in Yourself as a Leader

As the leader of your organisation, you have to trust yourself to make the right decisions in a highly dynamic environment. This trust stems from an honest conversation with yourself, regarding your own ability to keep up with the technology pace and with your fiduciary responsibility towards your stakeholders. And when you are not happy with your own performance, it is your responsibility to make your exit gracefully and responsibly.

AI IN REGULATORY DESIGN

Regulators increasingly have to apply AI-supported analytical methods to recognise:

- vulnerability patterns
- scan lengthy reports
- analyse incoming data

AI models might be able to capture all these interconnections, which are increasingly complex in domestic regulations and are becoming extra complex when overlaid with the interconnectivity of other jurisdictions' regulations. In the aftermath of 2008, regulators aimed to minimise risks but what evidence suggests is that they actually shifted the risks to other areas where risks are even more cumbersome to control. It would seem that supervisors do not necessarily understand the implications of the laws and interconnections, which might increase the systemic risk in the markets.[296] This would represent a divergent direction from the global efforts to harmonise regulations globally for a more consistent systemic risk management. There are concerted efforts led by independent bodies like the FSB or other regulators like the UK's FCA or Singapore's MAS to create a global testing ground for financial innovation. The UK's Financial Conduct Authority (FCA) has joined forces with 11 other regulators to create The Global Financial Innovation Network (GFIN), which creates a global testing sandbox for financial innovation. AI is regarded as having a cross border impact and features an element of the emerging technologies/business model. The current UK regulations put the ultimate responsibility on banks, and the FCA has clearly indicated that the ultimate responsibility for AI projects is with the senior managers, including the boards.

RESEARCH & DEVELOPMENT CENTRES

Research and development has been widely embraced in large organisations. From RBC, with their top talent Borealis, to Fidelity Labs, organisations aim to innovate in the safe confines of internal testing labs. Incubating new business within the organisation generally has a framework that aims to:

- Disseminate innovation and ideas internally
- Teach concepts
- Test-drive new ideas and put them in production

From discussions with technology executives when they refer to innovation labs they do not call them "centres of excellence" but rather experimentation or testing labs. This is a key lesson financial services needs to learn. Excellence comes only after testing, failing, and repeating until it works. In financial services, testing labs are pompously called *Excellence Centres.* They operate in isolation in some pseudo ivory towers in the shape of a cool fintech space with visible pipes in the ceiling, beanbags, and the obligatory pool table. Their mission is important but the perception by their "non-excellent" fellow colleagues is detrimental to creating cohesive and supportive transformation teams. This descriptor may have been fuelled by leading research magazines, which referred to testing labs as *excellence centres,* perhaps because of a lack of a better choice of words.[297] In real life, transformation does not happen in *excellence centres* but on the front lines with clients, managing day-to-day real-life problems.

EXAMPLES OF QUESTIONS FOR BOARDS TO ASK

	TOPIC	POSSIBLE QUESTIONS
1	Business Strategy	1. Where do we want to be in 5, 10, 20 years? 2. What are our business goals and aspirations? 3. What is our customer-centric strategy ?
2	Data	1. What is our data strategy and information architecture? 2. What changes do we need to implement?
3	Data Processing	1. What is our data processing strategy? 2. How do we make sure that we extract insights and then act upon them?

4	Workforce	1. How do we address the skills gap from board down? 2. What is our strategy to attract and retain AI talent? 3. Have we set up a technology advisory committee? 4. How can we bring in AI savvy directors on the board?
5	Ethics	1. Do we have a process in place to address ethical considerations? 2. Who is our AI ethics expert in our advisory team?
6	Regulations	1. Is AI testing and deployment in line with regulatory requirements? 2. Have we updated our risk management?
7	Research and Development	1. What capabilities do we have to build an in-house R&D? 2. Can we partner with an academic centre/AI adviser?
8	AI Adviser	1. Have we identified any conflicts of interest? 2. Is our adviser impartial? What network do they have? 3. Are they impartial in AI vendor selection?
9	AI Implementation	1. Do we have an audit of all AI projects running? Any duplications? 2. What are the challenges identified ?

KEY POINTS TO REMEMBER

- Technology must follow the business strategy.
- Business transformation with AI goes deep into the fabric of a business with ramifications in human resources, infrastructure, and management style.
- The 10-step framework for adopting AI includes: educate your staff, reimagine your business, choose your AI adviser, upgrade your board, design your AI strategy, re-design your information architecture, identify your first AI project, select your AI provider, test your AI, and deploy and scale your AI.
- One way to handle internal resistance to change and the fear that drives it is to adopt AI tools to solve the staff's typical time-consuming tasks.
- Key benefits of AI adoption include: scalability, personalisation, efficacy, cost reduction, speed, accuracy, efficiency, and productivity.

- When choosing your AI adviser, it is more important to find someone who is independent with proven experience and knowledge than large consultancies with household names.
- AI strategy means reimagining your business with AI, so your business strategy becomes your AI strategy. The absence of a coherent AI strategy is a business risk. ROI with AI can take time–business transformation with AI requires patience and capital.
- Reimagining business models impacts the functioning of any organisation in at least four areas: business processes, products, people, and insights.
- The five main components of data strategy include: identify data and understand its meaning, store data in a structure and location that supports easy, shared access and processing, provide data to package and share according to the business units' needs, process data to transform it from a "raw ingredient into a finished good," and govern data.
- An ineffective IT strategy, dated infrastructure, failing IT, and a disjointed information architecture are the most significant risk management issues facing the board of directors.
- Workforce strategy may also need to change when adopting AI–it may require the creation of new positions within the company, reconfiguring jobs, and training employees on new skillsets.
- Boards would be well advised to ask hard questions about deployment of AI tools that allow surveillance and prediction of people's behaviour. Such practices are not only an assault on people's basic right to privacy but also when misused leave companies wide open to reputational risks that are hard to manage.

CHAPTER EIGHT

How AI Is Reshaping Financial Services

This chapter will provide a snapshot of the current trends and challenges of AI adoption in the three broad segments of financial services: investment banking, commercial banking, and asset and wealth management and insurance, with a quick look at credit unions, islamic finance, and central banking. This analysis is by no means exhaustive, but it will equip you with enough knowledge to conduct an informed discussion about what the industry is doing in 2019 and to further your reading.

So let's look at this chapter as a starting point to discuss how things need to change and look at how other, more forward-thinking organisations are deploying AI. This chapter discusses the main developments with AI in a few verticals and functions.

ASSET MANAGEMENT

In one of the talks at Davos in January 2019, the CEO of Asset and Wealth management at JP Morgan, Mary Callahan Erdoes, stresses an operational perspective in their business. JP Morgan has 50,000 technologists and "it is really hard to imagine not being a bank at scale." She emphasises that their job is "to help the consumer. All the AI and Big Data we have aim to help the consumer to make a smarter decision, that's the goal we're all seeking to do here. If you do not have scale and the ability to do that, it's going to be a hard slog."[298]

This conversation highlights key questions that might have answers in AI solutions already available and used by some organisations in the following areas:

- operational scaling (operations)
- data size (operations)

Addressing these questions has a profound impact on the headcount. In 2017 Opimas LLC estimated that AI adoption would result in approximately 230,000 job cuts in financial firms worldwide by 2025, with the hardest hit area being asset management (with an estimated 90,000 job cuts), but my estimation is that the actual number will be significantly higher.[299] GIS-Liquid Strategies Group was managing $13 billion with 12 people in 2016. In 2017, Kensho, an AI company, was acquired by Standard & Poor's (S&P) for $550 million in the biggest AI acquisition to date. Kensho was founded in 2013, with the intention of replacing bond and equity analysts. Its algorithm is dubbed "Warren" (after Warren Buffet) and it can process 65 million question combinations by scanning over 90,000 events, such as economic reports, drug approvals, monetary policy changes, and political events and their impact on financial assets. Intel acquired Nervana Systems.[300]

Building an Ecosystem

Traditional asset management companies would need to develop their own platforms to access talent and new ideas. For instance, the hedge fund TwoSigma runs an annual Financial Modelling Challenge as a useful platform to access talent and ideas. Mindle AI is one of the winners of the 2018 competition, and they used sentiment analysis to identify market signals. There are also accelerators like BGI Lisbon Portugal founded by Goncalo Amorim. Such accelerators have a pedigree in deep tech that financial institutions can leverage.

Board Composition

The industry research and my direct experience demonstrates that this industry is experiencing the extreme of AI adoption–either all or nothing, enlightenment or ignorance. The latter category represents the majority and is driven by a rather deep-seated view, from the comfort of decent returns (the market worked in their favour for the past few years), healthy profit margins, and accommodating investors (everyone is happy when returns are handsome). This predicament is changing.

Starting in 2019, the strategy in many asset management houses will finally become the scrutiny of shareholders. Such scrutiny will stem from the eroded margins and weak financial results. Change is slow until it isn't, as in the case of a prominent example in the UK, Jupiter Asset Management, a listed company with $50 billion assets under management. They have historically been the choice of many UK savers. Their board composition is due to be changed this year, and in January 2019 they appointed a new CEO as their share price dropped by a dramatic 50 percent in 2018 on the back of dramatic outflows from their flagship fund Dynamic Bond. The company is on the short sellers list, at the time of writing this section. The technology investments in AI have been of secondary importance to the current board and the former CEO. This is no surprise. Jupiter's Board is currently composed of nine members and none of them have any technology background. They all describe their decades of fund management experience, but this is gradually becoming of secondary importance, as the organisation needs to compete in a data-driven market and because competitors are looking to access new business models (consider how BlackRock is exploring entering US wealth management by bringing technology-based investment and planning solutions through a partnership with Microsoft announced in December 2018).[301]

With such a background, it is naturally expected that Jupiter's board will do what they have done for decades in fund management. The organisation they run in 2019 is actually operating in technology business with a focus on fund management. So, technology skills, experience, and knowledge need to prevail. Anything else will place the company in an even more precarious position in the future. It would be interesting to see how the composition of the new board will change in order to bring technology literate board members. [302]

New Offering and Business Models

In March 2019, Goldman Sachs Group Inc. launched five new exchange-traded funds that rely on indexes built with machine learning and have the following objectives: (1) data-driven world; (2) finance reimagined; (3) human evolution; (4) manufacturing reimagined; and (5) new age consumers. GS is working with Motif Investing Inc. to identify companies operating in emerging technologies like artificial intelligence, genomics, and blockchain that will shape the next wave of innovation.

BlackRock has moved into providing administration to the fund-management industry with their product Aladdin, a risk analytics tool with comprehensive portfolio management, trading, and operations on one single platform. More than 85 fund managers are running $18 trillion in assets on the Aladdin platform. Through their partnership with Microsoft, they are attempting to enter wealth management and diversify their revenue stream. Aladdin is now tailored for wealth management, with UBS being the first wealth manager in the US to offer Aladdin to its advisors (discretionary investment advisers) who are able to analyse multi-asset portfolios. BlackRock's CEO Larry Fink acknowledged that by 2022 Aladdin will account for 30 percent of BlackRock's revenue.[303]

Around 2017 Amazon Web Services (AWS) started a systematic and low-key approach of asset management in London. The AWS approach seems limited to a very attractive cloud-based service for asset management. "You have to be crazy to turn it down, it's that good," a COO explained to me with a lot of enthusiasm. Well, I think you have to be cautious. There are many people who envisage investment management as a no-human-intervention process. So when AWS offers attractive deals, I would caution my reader to see beyond the obvious.

Profitability and Transparency

The FCA, the UK's financial services regulator, conducted a two-year consultation on the UK's asset management industry.[304] They took aim at the lack of transparency and price competition and recommended a range of new rules and regulations. They also found "high levels of profitability, with average profit margins of 36% for the firms we sampled. Firms' own evidence to us also suggested they do not typically lower prices to win new business. These factors combined indicate that price competition is not working as effectively as it could be," the report concludes. In 2019, the Woodford Investment Management meltdown is taking place as I am writing this section. It should unravel a range of conflicts of interest sheltered by the lack of transparency, and put under further scrutiny the profitability and lack of transparency in this industry. In turn, it will place further pressure on management teams to deliver returns and on asset management institutions to retain profit margins. Neither of these objectives will be reached unless the asset management house uses AI as a tool for business growth and expansion of new revenue streams. Although still substantial in 2018, the profit margins have systematically eroded over the last decade because of two main drivers:

1. regulatory requirements
2. volume and velocity of data

Regulatory requirements in the UK have placed a burden on the industry either in business process upgrades–primarily in technology investments–or fines. Regulatory requirements for transparency will continue to take their toll on profit margins. As these requirements become more complex, naturally the likelihood of breaching them increases. However, the volume and velocity of data is a more substantial long-term concern. In order to manage this concern, and satisfy the regulatory requirements, the majority of executives will continue to increase their headcount. This is an increasing cost on the balance sheet. The more innovative boards do not go far enough beyond mandating a mere tinkering around the edge with "digital innovation" by upgrading their current software and basically doing a patchwork to address inefficiencies in various business processes.

Global Trends in Fund Management

Quigdao is one of the first seaports in China open to trade with the rest of the world and is one of the most important coasts in China, with a beautiful historical and cultural legacy. It also ranks as 29th on the list of global financial centers and is moving up the ranks partly because of Quigdao's ambition to be the place where the fund management industry is being redefined. This is where revolutionary and innovative ideas take shape and are prepared to be rolled out. It hosts a number of leading conferences, such as The China Wealth Forum and The China Venture Capital Conference, and a large number of specialist training courses. More than 800 financial companies have settled in a dedicated state-of-the-art financial centre of 23.7 square kilometres, which is double the size of the city of London. More than 34 foreign financial institutions have also settled here, like Deutsche Bank and DBS Bank. In partnership with Chartered Institute for Securities and Investments (CISI UK), they provide quality training. Quigdao is looking to deepen their relationship with the UK, Switzerland, and Singapore.[305] With access to some of the most advanced and innovative approaches in AI, it is fair to expect that Quigdao will become a global asset management centre.

WEALTH MANAGEMENT

AI in wealth management provides a key delivery that this industry has never been able to reach: personalisation at scale in real time. It provides a host of opportunities to embed better behaviour by design,[306] irrespective of whether this service is delivered through financial advisers, discretionary wealth managers, or through DIY portfolios on platforms like Hargreaves Lansdown in the UK or Raisin in Germany, which in 2019 received further investment from Goldman Sachs. "Robo-advisors," a term that is misconstrued for what it describes (a better use of words would be "digital wealth managers"), such as Betterment and Wealthfront, have redefined portfolio management for wealthy, digital-savvy investors and show that "doing things the old way" is not an option in this sector.[307]

According to CapGemini's 23rd Annual World Wealth Report, in 2019, while high net worth individuals (HNWI) still place a great importance on personal connection with their financial adviser, nonetheless they value the digital engagement, and only 40 percent were satisfied with personalisation from their wealth managers. Service quality is the first criteria HNWI select in their primary wealth manager. This shows that AI is a huge opportunity still misunderstood and undervalued in wealth management. The report identifies a resistance to AI/digital advice adoption centered on the needs of financial advisers because (1) there is a gap between what the financial advisors see as their needs and their organisation's strategic needs, and (2) because of the organisation's processes and support structure for financial advisers. AI can deliver significant value in enabling wealth managers to better meet not just HNWI clients' needs but the needs of everyone. For instance, PIMFA (formerly the UK's Wealth Management Association) is driving increasing focus on the applicability of wealth management and financial advice to all individuals with robust programs to use AI to help people save and invest for their future.

It is not only digital delivery that is upending this sector. There is a new breed of offering in wealth management that leverages technology, and therefore it is able to offer exceptionally competitive rates. Fees are the core revenue of this sector. The race to offer low-cost products is termed the "Vanguard effect," a trend initiated by the asset manager. This is expected to change how the industry competes, as it collides with the traditional revenue-generation model (management fees on assets under management). And in order to remain competitive, technology needs to be deployed. There is no other way. Charles Schwab's Schwab Intelligent Portfolios are automated investing with zero advisory fees and zero commissions. Zero seems to be the ultimate objective for investment products in this sector. In February 2019, Vanguard joined the race of providing low-cost investment products, by cutting the management fee to a tiny 0.03 percent for a range of its ETFs (Exchange Traded Funds), including its $63 billion Vanguard FTSE Emerging Markets ETF, thus it costs a mere $12 a year for $10,000 invested. Forty-three other Vanguard mutual funds are also reducing their annual management fee.[308]

This competition continues to attract further interest, and the media was quick to label it as a "price war."[309] In some cases it had an immediate impact on the market, with an adverse reaction on the share prices of those providers that seem slow to responding. When Fidelity Investment brought to market zero fee index funds for retail, BlackRock shares fell 4.6 percent on the news, which also affected adversely T.Rowe Price Group, Legg Mason, and Franklin Resources. This is a sight to behold, something that the wealth-management industry would have never envisaged 15 years ago during my career in wealth management:

- Fidelity Zero Total Market Index Fund (FZROX) 0.00% expense ratio
- Schwab Total Stock Market Index Fund (SWTSX) 0.03% expense ratio
- iShares Core S&P Total US Stock Mkt ETF (ITOT) 0.03% expense ratio
- SPDR Portfolio Total Stock Market ETF (SPTM) 0.03% expense ratio
- Vanguard Total Stock Market ETF (VTI) 0.04% expense ratio[310]

Truth be told, these numbers are the nightmare of any CFO of a wealth management company. As their revenue model thins, so does their chance to exist as a business. Anyone in wealth management who thinks that he or she is immune to this change is wrong–to his or her peril. There are interesting options to explore, tailored to wealth managers, such is the case of Cognitive Scale, an AI company based in Austin, Texas. They built a suite of products that help wealth managers increase client loyalty, boost advisers' productivity, and reduce risk.

Hybrid financial advice (human and digital) will dominate the financial advice delivery, which needs to have a human-centered approach. Digitally-enhanced advice within an ecosystem of scale with focus on the client experience is where the industry is headed. The cornerstone of this business model transformation is creating cost-smart, risk-smart, tax-smart platforms that are able to support the delivery of advice personalisation in real time and at scale.There will be a handful of people who will prefer the concierge service of a wealth manager, but they will have to pay for it and they will be in the minority. For those who are astute DIY investors, the Vanguard effect is a great opportunity to run a low-cost portfolio.

Investment Behaviour

My academic research sits at the intersection of neuroscience, artificial intelligence, and wealth management in order to interrogate how episodic memory informs how people save and invest. A range of outcomes have surfaced in my research, but one stands out as essential because it encapsulates one of the most sensitive aspects of human existence: People's relationship with their money shapes their behaviour towards their money.

Professor Renee Richardson Gosline of MIT studied how experiences that are technology-mediated encourage improved behaviours in people. This work focused on analysing what created user experience comfort. The report also identified that "technology is moving toward 'Zero User Interface,' the ultimate, invisible user interface that allows technology to collect data and anticipate needs without direct requests or user activity." This report is important to wealth management because it enables the industry to "better understand the factors that affect customer adoption and abdication of control to benevolent technology."[311] For further reading, I recommend *Frontiers of Financial Technology* by David Shrier and Sandy Pentland for an informative book on future commerce, digital banking, and prediction markets and beyond.

The work with AI needs to focus on the circumstances and context to better understand how these technology-mediated experiences lead to better behaviours. Once this aspect is well understood in practice it will open up a host of new business opportunities for this sector, including family offices. Bridgeweave is a London-based AI company that aims to begin a new narrative with next-generation products for wealth and asset management to drive revenues, customer acquisition, and engagement.

High-tech Together with High-touch

The 2019 EY wealth management report identifies that 25 percent of the clients surveyed prefer face-to-face interaction or phone calls as their primary method of engagement with their wealth manager. For financial advice, 42 percent expressed the preference for direct engagement. High-touch seems necessary during a period of change in personal circumstances (marriage, death, children, divorce), significant market volatility, and for risk-averse clients.

A 2019 report identified that the average adviser can only spend 8.8 hours per week meeting with clients while a Charles Schwab survey identified that 79 percent of millennials want to have the option to access human advice. Industry leaders like Vanguard and Fidelity Go now offer hybrid advice models, which include access to a human adviser for a nominal fee.[312] AI is a suitable platform to achieve higher productivity and quality of engagement for financial advisers, enabling them to prioritise client service. This is an excellent opportunity for wealth managers to reimagine their customer engagement. [313]

Wealth Management for the Elderly

An average 10,000 Americans turn 65 every day, according to the US Census Bureau. The average net worth of families headed by those aged 65 to 74 was $1.07 million in 2016, according to the Federal Reserve. Fintech firms are leveraging forms of AI to deliver tools tailored to help adult children manage and monitor parents' financial interests and well-being, from paying bills to monitoring bank accounts.

While these companies offer tools that are useful, they are also a great platform to develop direct contacts with the beneficiaries of large wealth transfers. Healthcare Planning has built a tool that provides an assessment of the elderly's decision-making capabilities, with remedial actions recommended when needed. Adults aged 70 to 79 are estimated to have lost an average of $43,000 in each reported case of financial abuse. EverSafe and True Link Financial aim to fight financial exploitation.

The EverSafe tool is able to monitor through its credit check function when elderly names and social security numbers are used. In the US Eversafe is offered via Fidelity Investments and Raymond James Financial.[314]

Strategic Advances in Wealth Management

Wealth managers and professional advisers are now able to access synthesised, analysed, and organised infinite data that they draw on to make informed financial recommendations and investment decisions. ForwardLane is an AI company that provides an application that is deployed in a wealth-management firm's infrastructure to help achieve this access.[315]

The wealth management space has become a push and pull for market share capture between banking institutions and digital wealth managers (known as robo-advisers, a term that I prefer to avoid using for it is misleading). In the US, digital advice products continue to be at the forefront of recent launches by global banking institutions. US Bank launched their offering backed by FutureAdvisor, which is owned by BlackRock. Fifth Third Bank also announced the launch of their digital advisor. While banks launch digital advice products, digital wealth managers like Wealthfront or Acorns and Stash launched a checking account and debit card option. SoFi, a lender, got into the digital advice space and is now also offering checking accounts with debit cards.

It is worth noting that the established digital wealth managers are expanding their financial planning tools, empowering people with the tools they need to achieve financial goals like buying a new home or saving for school fees, or managing their retirement accounts or indeed having a current account. Envestnet and Betterment are reaching out to their customer base with enhanced banking capabilities through strategic partnerships with traditional banks.

Account minimums have been a deterrent for many people who want to access financial advice. Five-hundred-thousand dollars is usually the baseline to be accepted as a client of any leading wealth manager. This minimum represents the breakeven point for a wealth manager. However, there is a large market of up to $500,000, which couldn't be serviced. AI tools are opening new business opportunities with personalised engagement, automated portfolio managers, and risk-management tools. Interestingly, in a move to increase transparency and access to investments, Fidelity has reduced their minimum investment to $0, enabling unrestricted access to their platform. Data aggregators will become high prizes. The prelude was the biggest M&A in wealthtech when Goldman Sachs purchased United Capital for $750 million, which gave them access to new HNWI segments and state-ofthe-art technology like FinLife CX digital platform.

The next step in wealth management is voice technology, supported by Natural Language Processing (NLP), a subset of AI. As for the future, the current chatbots can be upgraded to an intelligent tool that is able to apply affective computing. The majority of banks have already deployed the digital assistance of chatbots supported by a RPA technologies. They now need to use AI, which they should have and could have used from the beginning. Wealth Wizard, a UK-based advice company, is developing a talking digital financial advisor. JP Morgan has deployed an AI-powered digital assistant in their treasury service division.[316]

CONSUMER BANKING

In 2018,[317] EY surveyed more than 22,000 digitally active consumers across 20 markets in order to assess mass consumer adoption and significant traction of fintech services. They identified four key themes in consumer banking growth:

1. Benefits: mobility, access, speed, value, efficiency, inclusion, ease of use
2. New services + players are driving higher adoption
3. Customers of fintech prefer using digital channels + tech to manage their lifestyle
4. Adoption will gain momentum with an expected rate of 52 percent adoption globally

The report identified an impressive, rapid growth in adoption with clear signs that fintech has reached a tipping point in its adoption. Fintech is also the first step to total reformation of financial services, and consumer banking has been one of the main beneficiaries. Fintech delivery is crucially connected to the deployment of AI–if it's not, then it is not adding real value in the long term. Currently, we are in the early stages of a fintech evolution; the more data fintech firms acquire over time, the more necessary AI will become to their delivery.

Incumbent, long-established organisations must compete to stay in the game. The strategy to compete in this newly formed market will shape a new breed of winners and losers. In the UK, Second Payment Services Directive (PSD2) allows, subject to the FCA authorisation and customer consent, data aggregators or non-banking institutions to access customer's data, which banks hold with the use of Application Programme Interface (APIs). PSD2 is the regulation which allows the "open banking" project to take place. This is also described as the evolution in retail and consumer banking.

There are a number of non-EU jurisdictions looking to adopt "open banking." In 2018, Australia commenced formal talks to adopt the same standard. The UK's nine largest banks were required to adopt "open banking" by early 2018. APIs open what was the historic preserve of established financial institutions and now give the retail consumer the visibility to access for price comparison and switching services. While open banking is meant to benefit customers, it pushes banks to compete even harder. How can they win? Only through offering radical personalisation at scale, and this can only be achieved with AI. AliPay, founded in 2004, is a Chinese platform that services clients across different categories like wealth management, payment, credit scoring, and insurance. In 2018, they had 850 million active retail clients and 12 million business accounts. By their own admission, personalisation at scale with AI is the secret to their phenomenal growth.

As early as 2016, JP Morgan announced building an ecosystem with fintech companies like Zelle for consumer payments, Roostify for online mortgages, TrueCar for auto finance, and OnDeck Capital, which provides loans for small businesses. In 2019, Société Générale announced their move to the next level of their innovation strategy under their visionary Chief Innovation Officer, Claire Calméjean. They aim to strengthen their relationship with the global innovation ecosystem through partnerships and investments in order to enhance their open-banking strategy.

The need for change is clear. The ability to effect change is clearly necessary, and yet banks are slow in adopting transformative programs–they continue to work in silos, they obfuscate internal access to data, and territoriality and inside politics undermine a bigger vision for change. Even if change is supported by a strong mandate from the board, it does not occur fast enough. Add a layer of technology legacy that is unable to support modern AI applications and a board that does not have the fortitude and self-confidence to request an infrastructure overhaul, and you can see the impossible task of delivering modern banking services. 11:FS, a UK strategic consultancy, has launched its successful business with one slogan "digital banking is only 1% done." They are right.

Many bank executives in big institutions are visibly exasperated that change does not occur fast enough, and that somewhere down the value chain, change is deliberately stifled and obstructed. A rather entertaining, candid, and colourful explanation comes from a former banker who has left such a big institution frustrated by the slow change and internal politics: "In a big institution, there is no benefit to taking business risk and succeeding, whereas the cost of failure is losing the senior seat that allows you to collect rental income on the franchise. The best strategy is to encourage a charismatic colleague to lead the charge, then after he succeeds, leak a photo of him having an affair with his secretary, scoop up his business line, and take full credit for the new revenue."

Many banks have committed to building their own digital offerings. Some have succeeded in a spectacular fashion like Marcus by Goldman Sachs and some have just started like Deutsche Bank with its Fyrst Bank. Others are putting their faith in the hands of technology firms like Santander Bank, appointing IBM to deploy AI, and committing to pay

$700 million (just for starters). Finally, there are banks that do nothing and just wait to see which challenger bank they will end up buying. It is no secret that this is the hope the UK challenger banks have when they are building a mobile-banking offering–that one day one of the global banks will buy them out. Revolut is going though a range of growing pains and cutting corners, and now it wants to act as a grown-up player and has appointed a range of silver foxes recruited from the likes of Citi and Goldman Sachs. Monzo has just closed another financing round, which valued it at over $2 billion and recently announced that it is actually making £2 per each customer, as opposed to bleeding money on unicorn valuations, a paradox many are unable to grasp. Starling Bank is expanding in a steady, grown-up and responsible manner, with quality corporate governance and visionary management.

Board directors from other jurisdictions that are only recently starting to become digitally driven asked me sound questions about whether they should encourage their bank to build a mobile-banking offering separate from the main business or aim to embed it in the current business. These jurisdictions have the advantage of learning from the lessons and the pains experienced by the fintech trailblazers. What big institutions are missing out on is their ability to generate full value from the data they hold. Some do not see the opportunity; others are not able to seize it because their infrastructure legacy is holding them back. Thought Machine has built a next-generation core-banking platform and they are working with a number of banks to implement it. It has taken the mammoth task of creating a core-banking platform from a clean slate to be a cloud-native modern alternative to the legacy platforms banks have. Such an AI-blockchain integrated solution will be the type of solution banking needs: a total overhaul starting from ground zero to free banks up from the legacy issues.

Any tinkering around the edges is not a long-term investment and is just a waste of money and resources with no long-term results. Software providers like Thought Machine, which will be able to deliver end-to-end plug-and-play solutions for specialist business models, will be able to reinvent financial technology offerings.

PAYMENTS

The highly lucrative space of payments is changing at a very fast speed. I would like to highlight two main developments led by technology firms:

- On 25th March 2019, 12 years after launching the world's first smartphone–the iPhone–Apple officially marked its entry into the payments space, with a strategic partnership with Goldman Sachs for Apple card, supported by MasterCard and built into the Apple Wallet app on the iPhone, offering users the ability to manage their card right on the iPhone. Apple achieved in one move what challenger banks have been trying to for years: easy signup, no late fees, no annual fees, no international fees, no over limit fees, works worldwide in wallet app, 2 percent daily cash back, in-app chat for customer support, maps with merchant logos, and a clear focus on privacy: no CVC number, no expiration date. Machine learning is used to label transactions, merchants' names and locations, then automatically colour-coded and organised per categories. Two percent of each transaction goes towards Apple cash to be used towards the Apple card balance or send to friends and family in Messages. Finally, Apple card aims to provide among the lowest interest rates in the industry. They use biometrics to authenticate the user either with facial recognition or fingerprints, in addition to a uniquely generated code.[318]

- Cross border payments has always been historically clunky, lacking real-time transfer and incurring high foreign-exchange costs. Libra, Facebook's digital currency initiative, was made public in 2019 and has been widely talked about as the disruptor to the payments sector. At the time of writing this, the prospects of making it happen seem bleak. Libra was not pegged as a specific fiat currency, but because it will sit on a reserve composed of a number of fiat currencies, it will have price stability. But it brings a range of anti-money laundering considerations and the banking system is naturally keen to make sure that any payment provider is subjected to the same regulations as traditional banks. Libra's technical underpinning remains an unknown at the moment, so is the regulatory scrutiny's outcome, which may determine its future. Mastercard, which is a member of Libra Association, also works with a consortium of banks to deliver instant payments. It is likely that we'll see a lot of change in this space, and AI solutions will be right at the core of this transformation.

LENDING

Consumer lending has benefited from AI adoption, thus securing speed and scale. According to CB Insights, Goldman Sachs reached $10 billion in loans in eight months, deploying Marcus as an autonomous lending platform (no human intervention). This puts in perspective the speed of scaling and business growth that can be unlocked. In the UK, OakNorth, a business lender and a digital bank, has started using machine learning for credit decisioning. The bank has lent £1.5 billion to UK businesses without a single default, directly helping with the creation of more than 5,000 new homes and 4,000 new jobs. The bank reached cash flow break even in just eleven months, had repaid all its accrued debts before it turned two years old, and made £10.6 million profit in 2017– only its second full year of operations.[319] Amazon is using machine learning to offer Amazon loans on an invitation basis only to sellers who qualify based on various data points like account tenure or customer experience. Amazon SME lending program was launched in 2012, and they issued about $4 billion in loans.

A key component in lending is credit assessment. Machine learning has proven exceptionally accurate and has enabled breakthrough access for financial inclusion. In the business lending space, NAV is a business credit card and loans product recommendation company, backed by Goldman Sachs. It is not a lender, but it offers a solution for small business owners to leverage their credit and financial data and it uses a lender-neutral algorithm. Personal and business credit scores are the main reason businesses get denied funding. NAV allows access to both scores, personalised insights, and tools to build business credit, so business founders can get approved.[320] In Chapter 5, Corporate Governance and AI Adoption, I talked about the important work Fujitsu has done in building explainable AI for credit decision tools for SMEs. Lending without explainable AI is hard to envision. Each lending decision needs to be transparent and explainable. Cignify and Juvo address algorithmic lending in emerging markets in Latin America and Africa, respectively.

INSURANCE

Regulation, product complexity, compliance, and large balance sheets will always be a barrier to entry in insurance. In the traditional model, those who own the customer relationships will ultimately be the most successful. Machine learning is being deployed with great success in improving business goals and with great ease to pilot these models:

1. Risk Modelling
2. Underwriting
3. Claims Handling
4. Coverages
5. Distribution

With AI, the insurance sector is becoming a "predict, protect, and prevent" rather than "detect and repair" industry, and the customer's loyalty will be with those who offer a better price and user interface. IoT (internet of things) devices (video surveillance, smart home ecosystem devices like thermostats, locks, smoke alarms, windows, and doors), sensors (temperature, water, infrared, sound), and data are going to tear down many barriers in insurance and transportation. These innovations will rewrite the nature of protective cover through a cloud-based digital ecosystem. This will open new markets, new products, and new revenue streams and operational models.[321] This will change the face of this industry. The investments in InsurTech have fuelled the use of AI technologies and disruption to the traditional business model in five different directions:

1. Traditional auto insurance premiums are likely to be substantially reduced.
2. Labour costs will be reduced.
3. Predictions will improve.
4. There will be new insurance products for customers.
5. There will be increased speed of delivery of payouts.

Since 2016, the number of applied AI cases in insurance has registered an increase with an improved rate of prediction accuracy after deploying deep-learning models. For instance, when used in claim determination to decide fault, non-fault, and split liability of a claim at first notification of loss in almost real time based on (1) multi-layer neural network working on diverse structured data (2) a combined multi-branch deep network that learns from structured (standard multi-layer network that extracts data) and unstructured data (Long Short Term Memory network type). Another use case is in claim fraud detection where a Support Vector Machine-based technique is used to analyse the details of the claim and of policy holders and then distinguishes among types of planned claims: legitimate, opportunistic, and planned fraud.

Investment in Insurtech start-ups are booming. In 2018 alone, about $4.15 billion was invested globally. In 2019, Prudential acquired Assurance with a surprising price tag of $3.5 billion. Traditional insurers need to protect their market share from these start-ups but also address the evolving complexity of cyber risk, telematics and getting more involved in corporate venturing. A sector that will register slow progress and market opportunity is small business insurance, a market estimated at $100 billion in the US. Another area that will benefit from using AI is commercial insurance for independent contractors looking to access pay-as-you-go insurance products. Square, the merchant services aggregator and mobile payment company, used patented machine leaning to predict and fund a business need. Telematics insurance has redefined car insurance, and it fosters three new business models: (1) usage-based insurance (2) pay-as-you-drive (3) pay-how-you-drive. Personal insurance has benefitted from more data points to create a better customer experience and improve prediction about risk via:

- in-car monitoring devices
- wearable activity trackers
- high-quality customer-facing apps
- virtual chat-bots to support customer questions and needs
- smart sensors monitoring connected homes
- web/mobile customer account dashboards
- software-as-a-service that manages insurance coverage and payment

New distribution models are shaping up. In Asia, Grab partnered with China's Zhong An to offer insurance. Tesla entered insurance in 2019, with their own insurance offering for Tesla smart cars, starting with California-based cars only, but with a clear intention to expand globally. Mergers and acquisition activity is regarded as the path to innovation by traditional reinsurers. MunichRe advances IoT, and AllState used M&A to move into theft and device protection.

There is a clear interest from life insurers to move into financial wellness, which will take away some of the wealth management current market share. Digital home insurance uses innovative underwriting and data for home insurance, while it aims to blend in the mortgage applications. The home security camera market is closely linked to home insurance, as the video surveillance market with mobile capabilities is growing at a steady rate. Home insurance claims innovation is also recording an unprecedented level of advancement.[322]

Cyber risk is complex and adaptable, but it has patterns to it. Companies like Envelope Risk use AI to redefine cyber risk modelling. In order to achieve this, they are combining knowledge from traditional insurance expertise with cyber underwriting, risk management, reinsurance restructuring, insurance linked-securities, and investment banking to build AI simulation tools and data-driven predictions.

Ambient assisted living combines smart home protection and connected health services as the bridge to a new strand of insurance for elderly who choose to "age-in-place."[323] Another new stream that is emerging is home insurance with voluntary employee benefits. FitBits and Apple watches are wearables that employees are encouraged to wear in order for employers to draw insights to keep them safe and healthy. It does not take long to wonder if this has any longevity because it blatantly encroaches on people's privacy. According to numerous academic papers and studies, data cannot be fully anonymised. You can always draw inference from other data points to identify people. In the same vein, life insurance is enhanced with facial recognition. Lapetus, a US insure tech, provides customised predictive life events solutions for life insurers and reinsurers. They have facial recognition technology that evaluates common health conditions, such as smoking ailments. If you combine this data (Lapetus) with fitness data (Apple watch or Fitbit) then insurers would gain a granular view of how to price their life-insurance products. If you add spending patterns and transaction history to the whole mix, then you obtain the holy grail in life insurance. At a 2019 fintech event, one global bank representative was thinking that a great way to monetise their data would be to sell it to insurance companies.

In this newly emerging space of personal information gathering for convenience and insurance premium calculation, Discovery Bank might actually cross the line. They claim that they let the customer behaviour dictate the price for its offerings. The South African bank is looking at its 4.4 million customers' behaviours and tracks their habits, and it claims that it offers better deals to those who live healthier lives. The bank is owned by an insurance and financial services firm, the largest health insurer in South Africa–Discovery Health. So here we have a set up that makes questionable use of personal data.

Insurance companies in EMEA (Europe, the Middle East, and Africa) are much more likely to indicate reluctance to up-skill their workforce, one report shows. Twenty-four percent of these companies suggest they will make little or no investment in digital skills and education this year, against 13 percent of retail banks.[324]

Alternatively, one might argue that there is enough body of evidence to demonstrate that in some cases algorithms are better at making certain types of decisions. An example is automated underwriting, with a group of economists demonstrating that those underserved in mortgage lending might be better off with automated underwriting than with manual underwriting. They show that such an algorithm provides higher accuracy and substantial benefits to underserved consumers.[325] Fukoku Mutual Life Insurance, the Japanese insurer, uses IBM's Watson Explorer AI to calculate payouts.[326] They are also among the first insurers that openly aim, in due course, to achieve full autonomy of their payouts.[327]

SwissRe Institute thinks that its value chain cannot be disrupted by small AI companies that are able to automate only segments of the value chain rather than the entire value chain.[328]

An ineffective IT strategy and failing IT have become business critical risks facing the boards of most organisations, in particular in the insurance industry. Chief Risk Officers and senior risk professionals active in the insurance industry need to be particularly alert to the main threats and developments in the IT landscape specific to insurance. They need to measure and manage these risks.

In the insurance underwriting business, machine-learning algorithms can be trained on millions of consumer data items (age, job, marital status, etc.) and insurance results (has the person been in a car accident, etc.) The underlying trends, insights, and correlations that can be assessed with algorithms to identify trends in real time that might influence insuring also detect trends that might influence lending and insuring into the future (are more and more young people in a certain state getting in car accidents?). In the insurance industry, Cytora is using AI to make better risk assessments about their customers, leading to more accurate pricing and minimising claims.

Reinsurance is another area where machine learning is used to reduce risk in global calamities. Munich Re, founded in 1880, is a specialist insurer in this area. They provide issuance producers and services to insurance companies to help mitigate the risk in a large-scale crisis. They have used their data to anticipate such calamities. They use image classification to assess how severe the damage is and produce immediate estimates, in order to accelerate payouts and help insurers to help people get back on their feet after their lives are affected by such calamities. Munich Re describes their use of AI as a tool for business innovation. They have a clear strategy led by a Chief Data Officer in (1) data engineering and storage, (2) analytics design of sophisticated analytics, and (3) AI used to solve essential problems with unstructured data in image classification and text analytics.[329]

PRIVATE EQUITY

An example of a long-term tactical move to secure new revenue models and also a leading position in the market, BlackRock, through their venture arm, aims to buy eFront. This is a strategic move as BlackRock seeks to become a bigger provider of the technology used by Wall Street's private markets. eFront is the world's leading alternative investment management software and solutions provider, and they use a range of machine-learning tools to deliver solutions for the private equity sector in performance evaluation, fund-raising, deal sourcing, data collection and portfolio monitors, and fund administration, just to name a few core functions of the private equity sector. eFront services clients worldwide and already has more than 850 clients across 48 countries. Their work also provides insights into improving the investment processes in private markets, naturally using machine learning to crunch large datasets that investment teams would never have the ability to do in order to draw valuable insights for improved investment decisions.

In private markets, institutional investors looking to invest in private equity funds strive to predict and select the top quartile funds. Between 1980 and 2012, the PE funds in the top quartile have produced returns over the public markets.[330] Research shows as much as a 16.9 percentage point differential impact on fund returns between being invested in a top quartile and bottom quartile PE funds.[331] A 2016 study[332] analysed 12,000 fund investments made by 630 limited partners (LPs) and concluded that an investor's fund selection is a more important driver than the mythical luck or access to top managers. eVestment, a Nasdaq company, is a marketing intelligence and performance analytics company. Their eVestment Private Markets is a tool that leverages a range of quantitative data to help institutional investors improve their selection process by making their quantitative diligence more efficient and thorough. While this is not an adaptive intelligence tool, it is a fair assumption that with the technology Nasdaq has and the volume of data eVestment collects from fund managers, their product will evolve and will provide more much-needed transparency in private capital markets.

FINANCIAL MARKETS

This space has benefited from advancements in AI at its various early stages. And yet, the march of automation and optimisation continues and transforms how this sector works. Goldman Sachs shows just how transformative automation can be to trading. In 2000, its US equities trading desk in New York employed 600 traders. Today, that desk has two equity traders, with machines doing the rest.[333] "In 10 years, Goldman Sachs will be significantly smaller by head count than it is today," Daniel Nadler, CEO of Kensho, told *The New York Times*.[334] Expect the same to happen on every trading floor at every major financial organisation. We already see signs of these readjustments with Deutsche Bank and JP Morgan reporting significant layoffs and divestments.

Trading is a closely guarded activity and providers aim to build their own proprietary tools to retain their edge. The Dutch banking services provider ING was founded in Amsterdam in 1991. In December 2017, ING launched a new tool called Katana, designed to help bond traders make better and faster pricing decisions.[335] Katana was developed by

ING's Financial Markets Global Credit Trading team in London. ING believes that the platform can learn from historical and real-time trading data and predict a statistical forecast that traders can use in addition to their natural intuition.

In his book *Advances in Machine Learning,* Dr. Marcos Lopez de Prado outlines the challenges in building an adaptive quantitative investment platform. The book also provides a range of solutions to many burning questions that quantitative finance practitioners have. This community has found many answers on open innovation and crowdsourcing in machine learning for quantitative trading, as it helps provide access to multidisciplinary research, which adds value to finance but also provides insights into what is popular in the ML community and ML systems. It might prove a valuable tactical approach for banks to build these crowdsourcing and open innovation labs.

Data visualisation is another avenue that is being explored in trading. Velocity Analytics covers the entire trading cycle as a high performance hub for implementing analytics, algorithms, and full-scale applications. The platform provides detailed financial and risk data at speed and scale and it is flexible in that users can develop applications and models faster with APIs for the adoption of tailored trading strategies. It provides data from 500+ venues going back to 1996 (corporate actions, tick history, reference and news data across asset classes while providing one single solution for compliance workflows).[336]

Citibank's aim is to collect and disseminate data centrally, so that they can not only use and interpret the data for trading purposes but also help their clients to manage risk–to identify unknown pockets of risks and manage the known ones. Their goal is to provide visualisation of millions of data points, and they have built a Proof of Concept (POC) using HoloLens. Data points used in real time and sentiment analysis from unstructured data (social media, articles, and videos) potentially make better trading decisions. Data points are now expanding to use real-time satellite information to identify a company's physical activities,[337] with a view to improving the quality of the trading data. Real-time access to IOT data (sensors, satellites, etc.) is the next level of quality of data aggregation.

High frequency trading (HFT) is regarded as the "natural evolution" that has transformed the trading floor into electronic venues, a competing space where the speed of receiving information and speed of execution have been critical ever since the time when Paul Julius Reuters used pigeons to transmit important stock news from the Paris stock exchange to Brussels in 1850. In a sea of changes, regulatory requirements follow suit. Algorithmic trading, where orders are entered, modified, and cancelled by an algorithm, poses a new range of administrative risks. In high frequency trading when these changes occur in a very short time span–we are talking milliseconds and seconds here–it can overload the trading system's infrastructure. It can also have a domino effect in instigating market events as a network, triggering other algorithms, which in turn can create a false market volatility, algorithmic solution, and even anti-trust law issues in the regulatory space, driven by fundamental pieces of legislation like the Second Markets in Financial Instruments Directive (MiFID II).[338] Regulators have started addressing these events and the regular burden on investment firms will continue to grow and be ever more complex as a reflection of the complexity of their operating environment.

Given the type of trading activity, HFT is highly demanding on the trading infrastructure, as it aims to reduce latency patterns when executing orders. An interesting chapter to read by Randolph Roth, a board member of Eurex, and Wolfgang Eholzer, "The Role of High-Frequency Trading in Modern Financial Markets"[339] concludes that "In modern electronic markets, it is effectively not possible, to provide liquidity, without utilizing HFT technology. The reason is that liquidity providers generate quotes on a certain time-sensitive information basis." In my conversation with Randolph Roth, it was interesting to also find out that they use machine learning for predictive maintenance of trading infrastructure. The nature of trading venues is changing fast, driven by technology advances and fierce competition among the exchanges.

In this complex space, I would like to conclude with paraphrasing the Head of Electronic Execution at a global bank who told me that ultimately "quantitative trading ideas are still generally human thought experiments rather than Unsupervised Learning Algorithms." Too true!

CENTRAL BANKING, REGULATORS, AND POLICY MAKING

The core application that is adding value to the traditional approaches is agent-based modelling (ABM). ABM or multi-agent systems or systems for modelling behaviour are powered by AI. They are useful tools for analysing complex scenarios that involve multi ecosystems and input while analysing human and institutional behaviour. There are a number of initiatives that are worth noting: Causalens predicts scenarios in real time and Symudine offers central banks a simulation tool suited to identifying the most likely scenario.[340]

Data collection and correct validation of regulatory datasets is a constant concern for central banks, with plausibility checks and anomaly detection as essential features. The Bank of England has appointed MindBridge to help them "provide data visualisation and data preparation techniques for larger numeric and transaction-level datasets, including credit union datasets and a daily dataset of trades submitted for the calculation of the [...] benchmarks."[341] Vizor Software is the global leader in SupTech solutions for financial supervisors. Vizor Software is an integrated data collection and supervisory system that is based on best-in-class supervisory models but flexible enough to deal with regulator-specific needs. De La Rue is a leading creator of nations' banknotes that meet the individual needs of each currency and circulation environment, provide an analytics solution to help central banks better understand their cash cycle, and make evidence-based decisions.

Financial strength ratings (FSRs) are significant since the 2007 to 2009 crisis, when relying on rating agencies proved fatal, as they failed to forecast defaults and the imminent downgrade of other banks. Regulators now have the ML tools to help with this.[342]

AI can help regulators in their work and alleviate a range of pressures on more data points collection. Policy making needs to be mindful of how it will impact the diffusion of AI in the industry. It should also consider where a country's citizens' financial data should reside, and if the central banks need to have a clear direction to follow to protect this data.

SPECIAL CASE: CREDIT UNIONS

Credit unions are community-based financial co-operatives that provide savings, loans, and a range of services to their members who own and control them, so their business model is providing the best service to members and not maximising profits. Credit Unions are key players in financial services, but unfortunately they do not receive the coverage they deserve.

In Britain, the first credit union was registered in 1964. Currently credit unions serve 1.5 million customers. Two hundred and sixty million people are credit union members in 117 different countries, and their global umbrella is World Council of Credit Unions (WOCCU), which has made inroads in financial inclusion but also advocacy at an international level on behalf of their members. In Australia, credit unions' aggregate asset under management places them as the 5th largest financial "entity" after the four leading banks. In 2018, I was invited to keynote the Customer Owned Banking Association′s annual congress in Melbourne, Australia, and this exposure gave me direct insight into some of the challenges due to size and sector fragmentation:

- Most have small datasets, which precludes them from leveraging AI. Some have around 5,000 clients.

- How to leverage omni-channel interactions to retain members, while encouraging new members to join
- How AI plays a key role in future growth, and how it can encourage new member acquisition
- The risks that these changes bring, and how to effectively mitigate them
- They need to appoint one of the usual top technology consultancies, as it enables them to defer liability in case anything goes wrong

There's no doubt that sector fragmentation leads to fragmentation of resources and datasets. Ireland mutuals have joined forces to effect change and invest in the next generation of technology. I'd like to leave the reader with a clear understanding that small datasets can be solved with transfer learning. Transfer learning is an approach to machine learning (a subset of AI, see more in Chapter 2, Taxonomy of AI) that enables a high level of accuracy in small datasets with very little training time.

SPECIAL CASE: FINANCIAL INCLUSION

The increased application of AI technology in capital markets is likely to reduce barriers to entry for many individuals who might not have previously had access to financial markets. In addition to this, portability of identity would open up new opportunities for financial inclusion.

Currently, phone-based biometric identity is being used for unbanked and undocumented people. In Kenya, one of the users of these applications described the use of AI as making it all a human-like experience. When this digital identity is embedded in blockchain, it can include immutable births, health and education records, or property titles. The key is data-mining algorithms, which are used to assess credit worthiness. The Chinese WeChat App has recently announced that they have 20 million new users each month. Grab Financial Group focuses on small and medium lending as a means to empower financial inclusion and access to capital. They have 600,000 merchants in their ecosystem, and they have built the largest payment ecosystem in Southeast Asia.

Headquartered in Kuala Lumpur, Malaysia, Alliance for Financial Inclusion (AFI) is a global network of central banks and other financial regulatory institutions from emerging and developing economies and is comprised of about 100 member institutions from approximately 90 countries. It was founded in 2008, as a project funded by the Gates

Foundation, with the "goal of advancing the development of financial inclusion policy in developing and emerging economies."[343] AFI has a program with the University of Oxford focused on the inclusion of women using disruptive technologies.

SPECIAL USE CASE: ISLAMIC FINANCE

Islamic banking assets were at US $2.4 trillion in 2017, and major global banks see them reaching US $3.8 trillion by 2022, with fintech credited for part of this growth.[344] A rising interest in Shariah-compliant products to serve the Muslim population, which is estimated at 1.8 billion, is also associated with this growth. The majority of this interest has been registered in the emerging markets where the majority of the globe's Muslim population lives, and it overlaps with work done for financial inclusion.

Substantial investments are needed at the interface between Islamic finance and fintech, with the predicted benefit of improving access to finance, personalisation and improving financial inclusion. In 2015, an IMF report identified a substantial gap in financial inclusion in the Organisation of Islamic Cooperation (OIC) member states, where less than 30 percent of households have an account with a financial institution.[345] Thirty-four percent of Afghan adults and 27 percent of Iraqis and Tunisians cite religious reasons for not having a bank account. Shariah-compliant lenders are at a disadvantage, due to lack of scale and high administrative costs. AI tools have a positive impact on many aspects of offering Shariah-complaint products, including speed and accuracy but also in reducing costs in credit assessments of the unbanked, processing phone-based biometric applications, and providing customer service with conversational AI (virtual assistants). A Shariah-compliant digital wealth manager is a welcome innovation, with Wahed Invest (USA) launching two Shariah-complaint index-tracking funds. According to the IFN fintech, as of March 2019[346] there are 120 Shariah-compliant fintech companies, with the largest number based in Malaysia, followed by Britain, Indonesia, and the United States. Two-thirds are focused on payments, remittance, and crowdfunding. The General Council for Islamic Banks and Financial institutions (CIBAFI) is regarded as the official umbrella for Islamic financial institutions in the world and counts 120 members from 32 jurisdictions. According to CIBAFI some of the top concerns have been identified as (1) service quality, (2) business growth, (3) expansion, (4) consumer attraction and retention, and (5) Shariah standards, compliance, and governance.[347] CIBAFI identified AI as "top of the list to be watched for in the next three years."

EXAMPLES OF QUESTIONS FOR BOARDS TO ASK

	TOPIC	POSSIBLE QUESTIONS
1	Business Strategy	1. Do we have regular industry updates on how AI is affecting our business? 2. Do we need to increase the number of board strategy days we hold? 3. Have we decided where we want to position ourselves as leaders or laggards in adopting AI? 4. Are we bold enough to keep up with our competition? If not, what do we lack? 5. Have we considered the cost of not adopting AI at scale in our organisation?
2	New Models	1.Have we held a brainstorming day with a large majority of our staff, including those who are considered less qualified, to discuss ways in which we can reimagine our business models and create new revenue models?
3	Our Clients	1.Do we still send those repellently worded letters to our customers? 2.Are we focused more on competing in the market or on serving our clients better? 3.What is our unique value that we bring to our client?
4	Change with AI	1.On a scale from 1 to 5 how committed is our board to give a strong mandate for change to the executives for the strategy with AI? 2.On a scale from 1 to 5, how confident are we that our board has a clear view as to what emerging technologies offer to our business model?

KEY POINTS TO REMEMBER

- AI solutions in asset management include operational scaling and new uses of data sets.
- In wealth management AI delivers personalisation at scale in real time. Digital wealth managers have redefined portfolio management for wealthy, digital-savvy investors.
- Wealth management that leverages technology can offer exceptionally competitive rates.
- With the growth of open banking in consumer banking, banks must compete even harder and offer radical personalisation at scale, which can only be achieved with AI.
- AI adoption offers speed and scale in consumer lending. Machine learning can help in credit decisioning.

- Machine learning has been deployed in insurance in risk modelling, underwriting, claims handling, coverages, and distribution.
- In private equity, AI can offer insights from large datasets and help predict the top funds.
- Automation and optimisation is transforming financial markets, with machines replacing equity traders. Some AI can help bond traders make better and faster pricing decisions.
- Data visualisation is also being explored in trading, providing detailed financial and risk data at speed and scale.
- Credit unions could benefit from transfer learning, which does not require large datasets.

CHAPTER NINE

AI Applications

In this chapter, we'll look at use cases where AI is being implemented in the areas of customer service, portfolio management, sales and marketing, personnel, legal, risk management and compliance, fraud prevention, finance, and IT. Whether it's projecting customer revenue or analysing customer service calls for better retention, AI can have a profound impact on client relationships. AI is currently being used in a large percentage of equity trading, and access to real-time data can positively influence buy-and-sell decisions for stocks. When marketing your business to new customers, AI tools enable greater personalisation, tailoring content specifically to users to avoid content fatigue. Personnel departments are also finding AI valuable when screening and engaging candidates and they have begun using chatbots for recruiting purposes and deploying more personalised learning in training and professional development. Rather usefully, in risk management, AI can be applied to help identify insider trading or cases of fraud when onboarding new clients.

This chapter showcases a number of business processes that are delivered with AI technologies. The use cases discussed do not represent an exhaustive list of what AI technologies can deliver. They shouldn't be seen as potential stand-alone "innovation" projects, rather they should be regarded as components of a coherent AI strategy.

The AI strategy is a joined-up approach to AI deployment in order to support the organisation's business goals, growth, and profitability. The key pre-requisite to a successful deployment is to correctly identify and then break down the tasks that can be automated. It is surprising how many organisations struggle to clearly define tasks' interconnectivity (process workflow) at the: (1) business processes level, (2) department level, and (3) enterprise level. Without a clearly understood process workflow, AI adoption's impact can be disjointed, nil, or chaotic.

For instance, a global bank realised that they were spending too much on AI projects with very little business impact, so they brought Cognitive Finance Group in to run an audit of their "AI projects." The audit identified an 80 percent duplication rate of projects with a 90 percent disconnect from the enterprise-wide workflow and 60 percent focus confined within departments. This is why it's critical to take a comprehensive AI approach that considers the growth of the entire business as a whole.

CUSTOMER SERVICE

Machine learning offers tools to spot consumer trends before they occur. Such tools offer sophisticated customer insights, such as dynamic customer segmentation, the lifetime value of a customer, analysis of customer service calls, customer survey sentiment analysis, predictive customer service, real-time personalisation of services, AI assistants, and personalised promotional planning. We'll take a more in-depth view of each of these areas below.

Adaptable or Dynamic Customer Segmentation

Based on behaviour patterns and preferences, machine-learning applications allow visibility across customer engagement behaviour and find new business opportunities by keeping customer lists that you might not envisage. For instance, it is easy to obtain a list of customers who got a mortgage and did not get home insurance, or customers who got a personal credit card and a business credit card. Deep insights into customer purchase behaviour can provide, for instance, explanations as to why some customers never get a second credit card or as to why some customers are likely to leave your business. Predicting customer churn or drop off would enable the business to adjust its offer. While these techniques are frequently used in other sectors, the retail financial services sector seems slow to see the value in deploying it.

Neo-banks like Starling Bank (UK) or Atom (UK) and neo-insurance companies like Lemonade (US), GetSafe (Germany) and CommerzVentures (Germany)are moving steadily towards embracing such tools to deliver an exclusively digital and personalised experience. While neo-insurers may not take risks on their books, they essentially replace brokers' appointments and provide seamless digital admin for quoting, binding, and issuing of policies, document and proof of insurance, electronic billing and payment, and real-time policy management. PegaSystems, a Boston, Massachusetts, company, delivers award-winning capabilities in customer relationship management (CRM) and digital process automation (DPA), powered by advanced artificial intelligence and robotic automation. They have been used by a range of clients from JPMorgan to OCBC Bank, American Express to ANZ, to manage their customer needs in consumer banking, wealth management, and investment banking. Cognitive Scale, based Austin, Texas, pairs human and machine intelligence to augment and enhance customer experience with trusted AI; their ethics is rooted in deep moral values, which they embed into their technology.

In the context of prevalent intelligence assistance, Professor Renee Richardson Gosline of MIT's Initiative for Digital Economy, part of the MIT Sloane School of Management in Boston, Massachusetts, researches the "Zero User Interface"–the interface that collects and anticipates customer needs without direct requests or user activity, a sort of mind reader. Professor Gosline's work looks into "understanding the factors that affect customer adoption and abdication of control to benevolent technology. To this end, this first study examines what combination of active and passive interface makes customers more (or less) comfortable with technology-mediated management."

Customer Lifetime Value (CLV)

CLV is the aggregate revenue that one customer generates for an organisation, and it is one of the key metrics for marketers. AI is used to estimate future CLV with increasing accuracy based on historic CLV and customer behaviour. Estimated CLV indicates how much you need to spend and where, in order to retain current customers and attract new ones. AI helps to make improved decisions to nurture clients by offering them tailored services. This leads to customers purchasing additional products, agreeing to longer commitments, and referring your organisation to their friends and colleagues. Klaviyo is a useful tool to calculate CLV.

Analysis of Customer Service Calls

Customer service calls can be analysed using text-to-speech software to transcribe, analyse, and mine conversations to identify issues and point where companies can improve their service. This can provide invaluable insights into areas that may need work for customer retention.

Customer Survey Sentiment Analysis

This type of analysis can help financial institutions automate the process of analysing and gathering insights from surveys using natural language processing NLP (a subset of AI).

Lexalytics is a Boston-based company that claims users can integrate their software (Semantria) with Microsoft Excel and perform text or sentiment analysis for surveys and social media data captured in Excel spreadsheets to then categorise the feedback.

Predictive Customer Service

Predictive customer service solutions have been used by USAA since 2016. They use AI technology built by Saffron, a division of Intel. It combines 7,000 different data points and is able to predict with 88 percent accuracy details like how certain customers might contact USAA next and via what media (web, phone, or e-mail) and what type of products they would be likely to be interested in.[349]Personetics is an Israeli-based company. Their offering is delivered via Salesforce and Microsoft CRM suite and aims to help bank managers and sales agents with insights personalised to each of their clients.

Real-time Personalisation of Services to Customers

Machine learning can help with identifying micro-moments in customers' financial journey and suggesting the best way to approach them. Sentiment analysis, image recognition, and voice analysis are used to identify those moments. Optimizely offers adaptive personalisation of website pages across every customer touchpoint. DataSine uses AI for better customer experience and helps financial services to better understand its customers. Wealth management is a clear beneficiary of such application. BNP Paribas has started using it.[350]

AI Assistants for Real-time Engagement

Sometimes incorrectly referred to as chatbots, AI assistants have five levels of development. They are expected to ultimately evolve to level five as outlined below by O'Reilly Report[351].

Table 9.1 The Next Generation of AI Assistants

THE NEXT GENERATION OF AI ASSISTANTS		
Level 1	Notification assistant	sends text message notifications (e.g. when credit card usage is 90 percent or customer is about to get into overdraft)
Level 2	FAQ assistant	answers simple questions from pre-programmed answers
Level 3	Contextual assistant	understands and responds to questions subject to their context (e.g. different inputs)
Level 4	Personalised assistant	gets to know the user over time; learns behaviour patterns, remembers them, and applies them when contextually appropriate
Level 5	Autonomous organisation of assistants	from lead generation to marketing, sales, or finance, these assistants will interact with each other within the organisation and will know each customer personally–this is not an unconceivable future

So far, we see levels one to three of AI assistants in use. Google Duplex might qualify straight to level four. The technology behind the Google Assistant is a complex autonomous system that places calls on customers' behalf, completed with a human-like voice, which understands complex sentences and context and is able to coordinate with customers' calendars. While this starts with simple appointments, the technology will be able to evolve and there would be applications in customer service centres. In 2016 Bank of America presented Erica, a virtual assistant, which is available at any time and performs anything from simple transactions to providing smart recommendations to their retail clients. Kasisto provides conversational AI solutions for the financial services industry. Kasisto built KAI–their conversational AI specialised in the language of finance–using decades of artificial intelligence research and IP from one of the world's largest independent R&D organisations that built Siri, SRI International. Kasisto is used by more banks around the world than any other conversational AI platform; customers include DBS, TD Bank, Manulife, JP Morgan Chase, Mastercard, Emirates NBD, and Standard Chartered, among others. Every day, nearly 18 million users have access to KAI through mobile, web, or voice channels. Banks are seeing up to 90 percent of their customer interactions handled by KAI without requiring any human intervention, with a 50 percent reduction in live chats, and four times more digital engagement.[352] Abaka is a London-based start-up that has developed AVA, a conversational AI with impact in savings patterns, which provides accessible and affordable advice on pension, savings, and investments. The company has a highly capable CEO and with a recent funding round in 2019, Abaka is likely to make their mark.

Personalised Promotional Planning

Rubikloud, a Toronto-based full stack, machine-learning platform, automates and improves mass promotional planning and loyalty-driven marketing for multi-billion-dollar retailers. While not yet used in financial services, it is a platform that would enable personalisation of promotional items.

PORTFOLIO MANAGEMENT

Portfolio management has been one of the first areas where AI was adopted. Autonomous learning investment strategies and sentiment analysis already have a notable impact on investment management. AI can play an important role in operations, accelerating change in the funds industry.

Investment Management Process

In one of their 2019 investment reports,[353] JP Morgan advises that "investment managers who adopt and learn about new datasets and methods of analysis will likely have an edge." They further suggest that any decision maker in the investment management function needs "to be familiar with big data and machine learning developments and related trading strategies." In their view, investors (fundamentals and quantitative) will leverage these techniques across all asset classes. In their report they conclude that human intuition is not replaceable by machine-learning algorithms so these are complex, long-term trends.

Machine learning can be used to analyse a multitude of data points, like capital market signals can be used to improve performance by reducing return volatility, drawdown, and correlation. Ainstein is a versatile AI tool developed by StockSmart[354] in MIT which helps investors, financial advisers and the boards to analyse and visualise in real time the impact of investment decisions.

Choosing the Right AI

The fund management sector has large datasets and is able to access substantial computational resources. The industry is quickly going to adopt AI to modernise their processes. In basic terms there are two approaches to machine learning in fund management:

1. black-box approach used by technology companies, with little transparency and more often than not biased recommending systems
2. causality approach applied in AI research circles like Lawrence Berkeley National Laboratory355 that uses an AI model to identify and rank many causes in real-life problems without time-sequenced data.[356]

"They use ML for building financial black boxes not financial theories, a data-mining exercise that ultimately sets investors up for disappointment," warns Marcos Lopez de Prado, author of *Advances in Financial Machine Learning,* a book that I highly recommend to board directors of asset managers and investment managers alike.

Equity Trading Strategies

Some reports estimate that today more than 90 percent of trading is being conducted by computer programs.[357] The industry is building their own AI research capabilities. Bridgewater Associates $150 billion asset under management (AUM) is running their own AI unit led by the legendary David Ferruci, formerly leading Watson at IBM. Renaissance

Technologies $65 billion AUM admittedly has the best physics and mathematics department in the world. TwoSigma Investments has put machine learning at the core of their investment strategy, with an MIT PhD in AI and an international mathematical olympiad Silver medalist on staff.[358] Sentient Technologies refer to their AI as "Evolutionary Intelligence" that creates trillions of solutions unique to a problem which then are tested using real data and rank the best performant genes. It is a natural selection of the best solution.[359] For those readers who are interested in more technical granularity, here is a list of applications of classification machine-learning algorithms in investment management:[360]

- Classification within context of alternative data
- Classification within context of industry, sectors
- Text and sentiment analysis

Applications of network analysis and clustering machine-learning algorithms in investment management:

- Traditional portfolio optimization based on correlations
- Clustering-based portfolio optimization
- Clustering-based portfolio diversification

Applications of forecasting and prediction machine-learning algorithms in investment management:

- Forecasting of financial time series
- Forecasting of risk premia
- Prediction of structural breaks: market regimes
- Prediction of structural breaks: bubbles and crashes
- Predicting company fundamentals

Challenges of machine-learning algorithms within the context of investment and wealth management:

- Not enough data
- Interpretability
- Finding appropriate parameters for machine-learning algorithms
- Overfitting

Autonomous Learning Investment Strategies (ALIS)

Investment specialists believe that we are at the cusp of the third wave of investment management, which hinges on the intersection of big data, data science strategy, machine learning, and computing power. The first wave was fundamental investing, the second computational finance. This is professed to upend investment management. ALIS managers are expected to take returns and alpha from large-capitalisation, fundamental, discretionary, long-short managers who will be left with investing in small and medium capitalisation securities, credit, and other capacity constrained strategies.[361]

Table 9.2 Academic Papers on Machine Learning in Investment Management

There has been a range of academic papers on deploying machine learning in investment management. Below is a list of papers that I recommend for those readers who'd like to examine machine learning in this field in further detail:

1. Improving Factor-Based Quantitative Investing by Forecasting Company Fundamentals by John Alberg and Zachary C. Lipton
2. Deep Learning for Forecasting Stock Returns in the Cross-Section by Masaya Abe and Hideki Nakayama
3. Forecasting ETFs with Machine Learning Algorithms by Jim Kyung-Soo Liew and Boris Mayster
4. Financial Series Prediction: Comparison Between Precision of Time Series Models and Machine Learning Methods by Xinyao Qian
5. Deep learning with long short-term memory networks for financial market predictions by Thomas Fischera and Christopher Krauss
6. Stock prediction using deep learning by Ritika Singh and Shashi Srivastava
7. An Artificial Neural Network-based Stock Trading System Using Technical Analysis and Big Data Framework by Omer Berat Sezer, A. Murat Ozbayoglu and Erdogan Dogdu
8. A deep learning framework for financial time series using stacked autoencoders and long-short term memory by Wei Bao, Jun Yue and Yulei Rao
9. Deep Learning and the Cross-Section of Expected Returns by Marcial Messmer
10. A Robust Predictive Model for Stock Price Forecasting by Jaydip Sen and Tamal Chaudhuri
11. Macroeconomic Indicator Forecasting with Deep Neural Networks by Thomas R. Cook and Aaron Smalter Hall
12. Stock market index prediction using artificial neural network by Amin Hedayati Moghaddama, Moein Hedayati Moghaddamb and Morteza Esfandyari
13. Recurrent Neural Networks in Forecasting S&P 500 Index by Samuel Edet
14. Forecasting Foreign Exchange Rate Movements with k-Nearest-Neighbour, Ridge Regression

Sentiment Analysis

A number of organisations have invested in sentiment-analysis tools that use AI to identify and predict customer sentiment in market movement. "The science of semantic disambiguation like sentiment analysis and adjudication veracity has evolved as algorithms ingest the spoken word," concludes SVP and Chief Data Scientist at Dun & Bradstreet, Anthony Scriffignano, who is passionate about the intricacies of linguistics and has deep experience in understanding data syntax and semantics.[362] In 2008, *The New York Times* reported on the precursor of sentiment-analysis tools in financial services, in an interview with Andrew Lo, the Director of the MIT Laboratory for Financial Engineering. They discovered a correlation between how after "anxiety" the word "bankrupt" appeared in the news and the S&P stop index fluctuations.[363] In 2019, JP Morgan built the Volfefe index, which measures Donald Trump's Twitter musing statistical impact on treasury yields. Citigroup also reports that these musings are becoming relevant in foreign-exchange moves.

Sentiment analysis vendors cover three main applications:[364]

1. Data search and discovery uses natural language processing applications. Alpha Sense offers market data collection. They have a search engine that is periodically indexed with millions of data points like public company filings and conference-call transcripts, and it also parses topics and concepts from these data points to find valuable correlations and investment information. With 800 clients and 128 employees, this New York AI has been successfully used in buy-side and sell-side to predict market sentiment and subsequent movements. The New York-based Dataminr offers a range of solutions to serve finance, corporate security, public sector, and PR. They focus on delivering real-time information on high impact events and critical breaking information that is crucial in real-time decision making. Another New York-based company, Amenity Analytics, founded by a former Wall Street analyst and portfolio manager, is used to detect sentiment, key commentary, and statements of apparent deception in the news, parsing for regulatory filings and earning calls for key points. They are used by Barclay, Citi, Nasdaq, Moody's, and other media companies and financial exchanges.
2. Report generation uses natural language generation and processing. Chicago-based Narrative Science is a leading provider of fraud detection reports like SAR (Suspicious Activity Reports), which financial institutions must file with the Financial Crimes Enforcement Network (FinCEN) after a fraud event has been detected or taken place. Narrative Science's Quill generates these reports. They claim that it saved a global organisation over $225,000 annually and reduced the time for

generating one report from 4 hours to 1.5 hours. The research departments would also benefit from using this report generation tool, especially on alternative data; the opportunities are abundant, especially as computing power and access to alternative data are becoming more accessible.

3. Process automation aims to use unstructured data and automate keyword identification, sentiment analysis, image recognition, and text summarisation capabilities and provide them as a package to the software users. Financial services organisations expect to receive on average 10,000 e-mails each day, which they then need to categorise and report to the regulator. Einstein Reply Recommendations, Salesforce Service Cloud agent, instantly suggests responses over chat and messages. Google's AutoML Natural Language is available through its Cloud AutoML platform. Eigen, it's Goldman Sachs-backed company, specialises in automating the classification and extraction of quality data from documents and other text sources of unstructured data. It uses AI, namely natural language processing.

Operations

Change is accelerating in the fund management industry, with increasing investor demand for speed and transparency. A solution has been brought to the market by FundRecs. Their Velocity platform covers trade capture, cash reconciliation, position reconciliation, and FX reconciliation. Their data processing and smart analytics enable users to take full control of their data without concerns about updates and lengthy installation times. Tora is a front-to-back provider for the buy-side. Their solutions offer portfolio, risk, order, and execution management systems and compliance and analytics engines.

UME, a Luxembourg-based company, aims to reduce by 90 percent the time management that companies and global distributors spend on due diligence. According to them, on average there are six intermediary layers between fund management and the end investors, which obscures the network hierarchy. Their data points are the foundation for future full automation of the exceptionally complex due diligence process in fund management distribution.

Symphony represents a challenger to the established Bloomberg service. Symphony provides not only competitive pricing but also enables the user to easily build chatbots that connect to their clients' ecosystem and trading platforms. Symphony was created when Prezo, a start-up messaging platform, was amalgamated with Goldman Sachs's internal chat system and raised $66 million from a consortium of 14 financial firms.[365]

Market Data Collection and Analysis

Illuminate Capital, a London VC founded by two former investment bankers, finds opportunities in data preparation in investment banking, as they are seeing adoption of data architecture solutions as foundational for machine-learning deployment. They focus on data access, standardisation, normalisation, single source of truth, and privacy. SteelEye, Privitar, and Ticksmith are examples of data architecture companies. Alternative data created a wave of excitement in 2017. Data like satellite imagery tracks car counts in parking lots, which would enable an inference on retailers or other correlations that otherwise wouldn't be immediate or that traditional measures might miss. However, there are a range of challenges embedded in the quality of this data, such as sample inconsistencies and a range of biases, which we discussed in detail in Chapter 5: Corporate Governance and AI Adoption. However, there is value in this data, when used with care and in economies with fewer data points that address direct questions about specific firms, industries, demographics, and geographies.[366]

AI data mining, text processing, and decision methods are used in the real-time trading data and news feeds to make automatic buy-and-sell decisions on stocks, commodities, and currencies. Up-to-the-minute news sources in digital form are readily available. The Reuters "NewsScope Archive" and DowJones "Elementized News feed" are used for automated trading and analysis. In 2019, the London Stock Exchange (LSE) agreed to buy the financial data company Refinitiv for $27 billion. If this acquisition is approved, it will turn the LSE into an important distributor of financial data and a top competitor of Bloomberg, the leader in the financial data market. This is the largest acquisition for the LSE, a market infrastructure company. There are other market infrastructure players looking to get into the data market.

Corporate Debt Trading

AI has already established a clear footprint in institutional income trading desks, and more advancements are expected, although at the moment most corporate debt transactions are still carried out via phone or messaging. UBS has tested an AI tool for pricing and trading bonds. ING's bond-trading tool is Katana. UBS is looking to phase out trade recommendations via their salespeople. Instead they are trialling recommendation algorithms in the corporate bond-trading division. The format is that of Netflix recommenders, which offers UBS clients a choice of likely interesting trades.[367]

Abbie is AllianceBernstein's virtual bond-trading assistant, which in its latest version is able to suggest junk bond trades to the investment group's fund managers.[368]

MarketAxess is a data-driven marketplace for trading bonds that counts over 1600 credit institutional participants. Their algorithms add a superior layer of liquidity for trading bonds. Trumid is a New York-based fintech that is expanding quickly in electronic bond trading. They bridge trading efficiency and market intelligence to support credit professionals with access to liquidity, within an intuitive design that fits into trader workflow.[369] Pagaya, a US-Israeli company, has a team of 20 data scientists and AI specialists and built a platform that analyses millions of data points to assess risk in different financial instruments. They use this information to select and buy individual loans. This is a departure from the traditional approach of securitising a pool of previously assembled assets.[370] In their words, they aim to achieve short-duration, high-yield return profiles with low correlation to the broader market and crisis resistant features.[371]

Nivaura, the London-based start-up, is an interesting proposition in capital markets. They leverage technology in a plug-and-play solution for digital investment banking services for the issuance and administration of instruments such as loans, bonds, and structured notes. Founded by Avtar Sehra, a PhD graduate in particle physics who spent a decade in capital markets, Nivaura brings notable value to modernising capital markets. Millennium Advisors uses AI to help trade off-market credit liquidity on over $1 billion a day.

SALES AND MARKETING

AI in sales and marketing is all about personalisation at scale. Emotional brand connection is cited as the main reason people stay with a service provider, when people feel "like they get me."[372] In exchange for this, according to a 2018 report, just under 80 percent are willing to share their data for personalised engagements.[373]

Too much data in too many places is one of the challenges of marketing. Apollo Insights is a platform that uses machine learning to identify opportunities and threats in the content of your marketing campaigns. They look at competitor intelligence, keyword search, technical Search Engine Optimisation (SEO) or content gap, and provide prioritised insights and actions that humans wouldn't be able to see in such a vast volume of data points. Apollo Insights algorithmically analyses three million websites, 120 million pages and 20 million words, and 1.8 million social profiles.

Content fatigue is a real stumbling block for content marketing engagement. Correctly tailoring content to users would enable a higher rate of permeability to new messages. Recombee is an app that has an adaptive recommendation engine, and switching to AI curation over editorial enabled a media company to achieve a reported 64 percent raise in conversion rates.

More often than not, financial services companies use a marketing agency to deliver most of their marketing needs. According to a CapGemini report,[374] 84 percent of marketing agencies plan to implement and expand their AI strategies in 2018 and beyond, 75 percent of enterprises using AI enhance customer satisfaction by more than 10 percent, and three in four organisations implementing AI increase sales of new products and services by 10 percent. LoopMe, a UK-based company with global offices, is part of this category.

Marketing departments have historically been treated as a cost centre rather than a revenue-generating centre merely because their contributions couldn't be measured. Thanks to AI tools, they can now measure the many contributions of this department to the business growth. They can now use machine learning to improve sales effectiveness, lead scoring accuracy, and trace it back to initial marketing campaigns and sales strategies. They can use use AI to optimise marketing campaigns, improve precision and profitability of pricing, and have visibility of the drivers of Sales Qualified Leads (SQLs) and Marketing Qualified Leads (MQLs). It is expected that by 2020 real-time personalised advertising across digital platforms and optimised message targeting accurate context and precision will increase.[375]

Machine learning can be used in customer retention efforts, analysing and reducing customer churn to streamline risk prediction and intervention models.[376]Machine-learning tools streamline revenue contribution for a diverse range of sales strategies by automating the entire process. TIBCO is a software company that enables better decisions and faster, smarter actions.[377]

Marketo is the most comprehensive ecosystem of marketing solutions that has been used in many asset management firms to support their marketing and sales capabilities (e.g. integration with CMS or CRM platforms, solutions for analytics and big data, and to build and manage campaigns from start to finish). Idio provides the asset management industry with an AI-driven solution that enables the organisation to personalise job and e-mail experiences at scale, without extra manual effort. E-mail campaigns are becoming more accurate in how they are designed and delivered, using some of the following tools:

- Optimail is a tool that makes sense of scattered analytics. The platform self-adjusts each campaign, depending on customers' past preferences and responses.
- Onespot helps with e-mail personalisation at scale, recommending predictive content on what a current or prospective customer might want to receive next.

- Phrasee analyses all current communication with a client and recommends exact wording that is likely to resonate with your customer. It's personalisation of language and style of writing.

CaliberMind, a US firm, uses a mix of machine learning and natural language processing to automate business-to-business (B2B) customer platforms. It allows marketing and operations teams to answer business questions and automate the customer journey from the unknown to happy customers.[378] For funds distribution teams such a B2B platform would bring insights and answer questions like: "Which marketing programs generate the most revenue?" "How is my marketing campaign performing?" "How fast are we driving revenue through the pipeline?" "How's our close rate trending over time?" The company claims that it has 10 times faster time-to-insights at 10 percent of the cost of traditional IT/BI solutions.[379]

Backed by Goldman Sachs, Data Fox is a company intelligence platform that uses AI to gather relevant data for sales and marketing and finance teams, basically updating and upgrading your static CRM database, putting an end to duplicate accounts, data anomalies, and stale data.[380] The updated data can then be embedded into the enterprise unique workflows for targeted results.

Wealth management, retail banking, and investment banking have found Pega a suitable solution for their needs. Pega's sales automation tool helps financial advisors to understand and analyse their clients' context in real time using AI that delivers proactive, relevant, and personalised next best action for new business.[381]

Finally, the Salesforce AI tool (Salesforce Einstein) has registered increased penetration and is now able to deliver personalisation at scale for each customer for an aggregate of one billion queries a day,[382] in different sectors, including financial services.

Marketing for attention in advertising is moving at a fast speed, and our industry needs to be aware of the latest progress. With a lot of privacy question marks and ethics queries for what is perceived as a questionable approach, cognitive adverts have arrived. IBM's Watson Ads use natural language processing to engage with consumers. It is an industry-first advertising product that uses Watson technology to help brands have "one-on-one personalized dialogue with consumers, at scale, and to deliver more relevant information than ever before in an advertising unit."[383]

As we march on using private data, a number of privacy concerns have been raised. People have become aware of their own data, and this awareness might change how AI solutions in marketing are being accessed and used. Meeco, an Australian company, provides individuals with data control for access and the ability to share personal data on their terms. Their solution is also applicable in financial services, and it is rooted in deep moral values that aim to restore data trust between people and organisations.

PERSONNEL

In the age of automation, when we need people to work alongside and supervise machines. The typical use cases of AI in the Personnel vertical include the following:[384]

- Candidate Screening helps with assessing the candidate quickly and effectively.
- Candidate Engagement provides feedback in real time, unique to the individual candidate–and not just driven by tags, positions, locations, or categories.
- Re-Engagement allows companies to re-engage and to update candidates' records to match new positions and update work experiences or skills.
- Post-Offer Acceptance follows up with the candidate in the period between leaving their previous jobs and starting the new one.
- New Hire On-boarding delivers the induction material, answers other common questions, and supports new hires as they adjust to their new job and specific requirements.
- Career Development provides personalised recommendations for career development programs or company coaching, which is especially useful in very large organisations where internal progression is desirable.
- Employee Relationships and Scheduling addresses employees′ engagement with the Personnel department, from addressing simple questions to scheduling a meeting to addressing in person more delicate questions.
- HR Compliance and Case Management documents engagements, incidents, and investigations.

A 2019 Gartner survey found that about 23 percent of organisations are using AI to manage their workforce. In January 2019, I asked a few top UK recruiters what intelligent applications they use. They didn't find any reliable solutions, meaning high accuracy rate. It shows that AI in this vertical needs some improvement. However, the recruiters did mention a few names. Humantic (formerly DeepSense) helps automate candidate assessment right at the top of the funnel by using AI with emotional intelligence to predict candidates' culture fit, personality, and behavioural attributes like teamwork or learning ability. For white-collar jobs' recruitment, a number of people were pleased with OnContractor, a match-making platform that uses prediction tools. Chatbots can also potentially help Personnel with candidate management and employee queries. More and more companies have started using chatbots to support their HR teams and the recruitment process. Jobpal, Wade & Wendy, Olivia by Paradox, Karen.AI, RoboRecruiter, and PeopleFirst by MHR are a few chatbots that have recorded a positive referral as of January 2019. Naturally, this is a dynamic space and many changes should be expected.

Other tasks delivered with AI tools are writing job descriptions (Textio), talent searches (Koru), screening resumes (Ideal), and on-boarding and training (Chorus). Avrio AI, a Boston-based firm backed by NXT Ventures, has built a recruitment platform that aims to combine all recruitment needs and to be useful to recruiters as well as candidates, in equal measure and 24/7. They offer a demo to anyone interested in checking out how their platform works. Tengai, the world's first robot designed to carry out unbiased job interviews, is being tested by Swedish recruiters. This raised a range of ethics questions, especially around the claim that it is unbiased. Tengai measures 41 centimetres (16 inches) tall and weighs 35 kilograms (77 pounds). She sits directly across from the candidate she's about to interview, on the table.

Headstart is a diversity-driven applicant-matching and management system. They use machine leaning to help clients find the right employee, regardless of gender, ethics status, or age. Accenture Lazard and Travelex are a few of their clients. They use technology deeply seated in moral values.[385]

Staff Surveillance

StatusToday, a London-based company, has created an controversial way of surveilling staff just to sell another product to surveillance-obsessed companies, adding another unnecessary layer of surveillance and privacy invasion. Their approach is to constantly track staff's meta data (file data, access card usage time spent in facilities, etc.) to build a normal behaviour profile. When this profile changes, for whatever reason–as we all have a bad day now and then–the system alerts the manager of "anomalies" in behaviour. Allegedly for a good cause, like avoiding employee burnout, but let's not gloss over the privacy invasion.

In 2018 IBM launched a questionable "employee retention" AI tool that can detect with 95 percent accuracy when employees want to quit.[386] IBM would not disclose how they do it. Cortexica is used in BT and Cisco. It uses image-recognition algorithms for health and safety checks. BT and Cisco want to ensure that their workers always wear the suitable safety protection equipment. Sybernetix (acquired by Nasdaq) warns when traders change their usual investment behaviour patterns like taking large positions in asset classes that they avoided in the past or have no experience in.

Training

KnowledgeOfficer builds a personalised learning journey to focus learning on key skills to secure the next area of progression. Their AI tool analyses job descriptions from the top companies in the world and picks the most important skills they seek and centers the learning journey on these skills.[387]

J.P. Morgan made a strategic investment in Volley.com, which uses AI to help large enterprises automatically generate training content for employees. Leena AI helps organisations solve training challenges for their staff and build a workforce with relevant skills and knowledge, in a conversation-driven format.

Personnel chatbots help with engagement level and create an interactive learning platform to facilitate training sessions. A personnel chatbot helps with follow-through commitment post training. This is an essential requirement for Continuous Professional Development (CPD) necessary in financial services. Chatbots can be easily integrated within the existent infrastructure.

Chatbots are be used to:

- gamify the learning experience to progress the learning adjusted to each user
- test application of knowledge, skills development, and information retention
- offer an insight into behaviour of staff and formulate tailored learning policies
- complete training at staff's convenience
- increase accessibility on mediums like smart phones with reduced dependence on classroom-style learning and in multiple languages.

Recruitment

These technological augmentation tools are useful, however, they must be envisaged as replacement of humans even for junior and trainee recruitment. Hireview uses computer vision to analyse footage of a candidate's interview and assess verbal skills, tone, and body language, in order to select the best candidate. Pymetrics evaluates candidates at the early stages of the interview process. The company claims that their product is agnostic to gender, ethnicity, and education level. Hiredscore provides feedback for rejected candidates, which might be useful. Eva.AI is a platform that claims that it can triple the productivity of a typical recruiter and shorten the interview time by 10 times and increase the candidate's satisfaction rate by five times. It can:

- source top talent using powerful prediction engines
- match people to jobs with precision
- screen every candidate at speed
- engage in an unlimited number of conversations at any time.[388]

Internal Communications and Workflow

Receptiviti, a Toronto-based technology company, applies AI and proprietary language psychology science to companies' internal communication systems, providing real-time insights into the psychological state of the workforce. This type of tool needs to be considered carefully before it is introduced in an organisation. They are surveillance tools over and above regulatory requirements. Zoom.ai aims to improve employees' work with their personal automated assistant, which learns individual personal preferences and is personalised to manage mundane tasks like searching for files, scheduling meetings, generating documents, and much more in over 170 languages.

LEGAL FUNCTION

"Will the future give birth to a new legal field called AI law?" wonders the authors of *The Future Computed,* a book published by Microsoft that I recommend.[389] This is a forward-thinking question and the answer is likely yes. As AI enters every layer of our society, from autonomous cars to algorithmic trading, from online advertisements to credit rating, from gig economy to employee surveillance, decisions are made that impact humans and that also define how AI agents compete with each other.

The authors continue: "Today AI law feels a lot like privacy law did in 1998. Some existing laws already apply to AI, especially tort and privacy law, and we are starting to see a few specific new regulations emerge, such as for driverless cars. By 2038, it's safe to assume that the situation will be different. Not only will there be AI lawyers practicing AI law, but these lawyers, and virtually all others, will rely on AI itself to assist them with their practice." This is a core signal that this profession will be redefined in how they do their work. This will have clear implications for lawyers working with financial services clients.

"The real question is not whether AI law will emerge, but how it can best come together and over what timeframe." I also agree with this vision. Law schools are now offering classes on data science. In order to remain relevant in their profession, some well-established lawyers in their mid-40s have gone back to school to take classes on those topics or even earn master's degrees. AI and law have become irreversibly connected. Law firms find themselves in front of an ever-growing volume of data that must be analysed and that can be done only using AI. Moreover, their clients demand more data-driven strategies from their law firms.[390]

In the legal profession, AI helps with a number of tasks:[391]

- Due diligence tools are used to uncover granular information from vast and unstructured datasets.
- Legal analytics tools use past case law, win/loss rates, and a judge's history to forecast the outcome of a case but also to identify behaviour patters and correlations that are not immediately obvious. In 2019, France outlawed these tools as they would appear to influence how judges decide.
- Document automation tools use templates and automatically create new documents based on specific data. NeotaLogic's tool shortens the non-disclosure agreement process by building personalised templates according to users' requirements.
- Intellectual property can benefit from the speed and accuracy of AI. Analysing large IP portfolios can be time consuming and errors can infiltrate with deep consequences. The accuracy of AI tools analysing such large volumes of work is likely superior to that of humans. TrademarkNow is a dedicated IP search engine. AnaquaStudio helps with patent application drafting for provisional patents. TurboPatent helps with patent applications review and also in creating legal claims.
- Laws firms can use electronic billing as an alternative to paper-based invoicing with Bright Flag and SmokeBall management tools, which automate recording of time and activities of lawyers. The full accuracy is not there yet with either of these tools, but it is progress in the majority of cases.

Like in the financial services space, legal firms have already entered into partnership with a number of legal AI companies, have invested in them, or built their own solutions. Labor and employment giant Ogletree Deakins Nash Smoak & Stewart announced a partnership with LegalMation, which automates early-stage litigation tasks using AI to analyse legal complaints and discovery requests, then generates draft answers and discovery requests, responses, and objections. Notably, Stanford and Suffolk law schools joined forces with the funding from Pew Charitable Trust with the vision to build a good legal NLP.[392]

Due diligence is time intensive and the value AI can deliver is measurable, especially as a Cass Business School study shows that thorough due diligence is linked to significant increase in deal success for acquirers. Kira Systems is a machine-learning company that aims to identify, extract, and analyse text in contracts and other legal documents. They claim that their system can deliver high volume contract data extraction in 20 to 90 percent less time *and* improve accuracy. Regulatory reporting and compliance reviews are areas that benefit significantly from such tools.

Document search, data discovery, and data mining remain the main time-consuming activities in financial services' legal departments. eBrevia (acquired by Donnelley Financial Solutions) provides automation in data extraction and contract analytics and claims that they are at least 10 percent more accurate than manual review and that they can analyse more than 50 documents in less than a minute. In 2017 JP Morgan built an in-house legal tech tool called COIN (Contract Intelligence) that interprets commercial-loan agreements, which previously consumed 360,000 hour of lawyers' time per annum. COIN can extract 150 attributes from 12,000 commercial credit agreement in seconds with high accuracy.[393]

ThoughtRiver, developed at Cambridge University, handles accurately detailed activities for improved risk-management functions like portfolio review and investigations, understanding context and summarising high-volume contract reviews. Ross Intelligence is a highly accurate search engine for lawyers. It can search billions of documents and provides the next contextual answer.

Luminance provides a powerful platform for instant insight for document reviews. They use machine learning for pattern recognition to identify similarities, differences, and anomalies at all levels of review, so that legal teams have a high level of confidence in their recommendations, irrespective of volume of data, jurisdictions, or number of teams working together.[394] They have a highly competent management team headed by CEO Emily Forges, are growing at a fast speed due to the quality of their product, and are being used by leading law firms and financial services firms across the globe. LawGeex and Eigen Technologies are two important AI companies in contract review that assist with compliance and risk-management functions. Brooklyn Law School, among other law schools, is part of a large consortium of US law schools doing computational AI. Further reading and useful updates are available from *artificiallawyer.com*.

RISK MANAGEMENT AND COMPLIANCE

The pace and impact of change across the financial system has fundamentally changed over the past decade. Globally, more than 33 percent of the digitally active population now use fintech services, more than double from 16 percent in 2015. Twenty-two million people in the UK regularly used banking apps during 2017, a 12 percent rise from the previous year. At the same time, the average bank branch received 104 visits per day in 2017, a 26 percent drop from 2012.[395]

Risk management has changed. The compliance teams have at least tripled in size in the past decade, and they have become a core cost centre and a determining voice in shaping the business strategy around regulatory requirements. "The role of risk is to enable business to do more, but in a controlled, safe, risk-aware manner, not just to stop risk taking," the non-executive director of a global investment bank wisely comments. The role of the Chief Risk Officer (CRO) has changed and has become a decision maker and adviser to the board, a challenger to business heads and a team player to the executive committee. As business decisions have become quantitative, the risk function works to synchronise business operations and business strategy, and it is also the medium that mitigates emerging risks.[396]

In financial services, as the industry is moving online, the risk function has become complex with moving targets. It will not get simpler. There are a number of AI applications that have made risk and compliance more efficient, as predictions have become more accurate by analysing vast volumes of unstructured data and identifying patterns and relationships in the data. The main areas where we find AI applications in financial services are in surveillance of conduct and market abuse. Trading violations that come with great regulatory penalties and reputational risks.[397] The first generation of machine-learning systems have been deployed in the following functions:

1. communications surveillance
2. archiving
3. mobile capture
4. voice analytics
5. voice recording

One of Goldman Sachs' portfolio investments is Digital Reasoning, a surveillance company. They have been joined by Barclays, BNP Paribas, Credit Suisse, and Macquarie Group, alongside other investors. Synthesis is their machine-learning platform, which understands human interactions, subtleties in communications, and undertones of context communication.

Pattern-based prediction is one approach to manage risk. Agent Based Modelling (ABM) is another emerging approach for risk simulation that uses AI to create scenarios with different variables and assumptions. Simudyne, a London-based start-up, runs advanced simulation trials/scenarios to reduce risk exposure. ABM, unlike regression-based models, does not rely on historical data. ABM looks promising in the risk function, especially as risks evolve in a dynamic and random fashion. ABM helps banks to test their decision before committing resources and taking any actions.[398]

Cube is a London-based regulatory intelligence platform, underpinned by machine learning and natural language processing technologies that reduce the operational costs of managing regulatory change.[399] Droit is backed by Goldman Sachs and uses AI to generate real-time information pre and post trade. Institutions have a comprehensive approach to regulatory requirements for decision making and execution.

Onfido, a London-based company specialising in government issued ID verification and facial biometrics verification, is used widely in financial services to prevent identity fraud during new client on-boarding. They cover 4,500 document types from 195 countries, addressing key KYC and AML regulatory requirements.[400]

FRAUD PREVENTION

Credit card fraud prevention is one of the first use cases in which machine learning has been applied for over a decade. The volume of historical data enables training, backtesting, and validation on vast datasets. The result is that transactions and responses occur in real time. In contrast, prevention of money laundering and terrorism financing though payments is still done using rule-based systems, which are not able to detect patterns and give a large number of false positives–they are not as accurate as the AI used in credit card fraud prevention.

Therefore, substantial human input is needed to identify and prevent suspect transactions. In addition to banks using outdated technology, the body of relevant data is limited due to data-sharing regulations and data usage. In this context Tookitaki aims to apply machine learning to fight money laundering. Singapore's United Overseas Bank (UOB) is their first client. From this partnership resulted a holistic machine-learning solution that would enable UOB faster access to more accurate insights to prevent and detect suspicious money-laundering activities.[401]

CitiBank invested in Feedzai, which uses ML to analyse omni-channel customer data from any source to turn it into risk scores and fight fraud in real time. Customer profiling helps with identifying and stopping unusual behaviour. Feedzai Genome is a dynamic visualisation engine that quickly identifies emerging financial crime patterns. Finally, their technology offers human readable results for users, analysts, and investigators, so they can understand the logic behind the risk scores, for more accurate decisioning. Their technology has proven superior in enabling organisations and their teams to fight financial crime dynamically. The executive management is a highly competent team lead by their CEO, Nuno Sebastiao. They have committed to a high level of ethics and explainability of their technology.

There are reportedly 90 voice fraud attacks every minute. The AI company Pindrop uses AI for voice identification. By listening to calls to figure out if it's really *you,* based on your voice print, behaviour, and the noise on the line, their technology is able to identify the fraudsters and keep people safe online.

FINANCE FUNCTION

Some of the machine-learning uses in the function of finance include:

- Budget Optimisation: top down from headquarters to each business unit
- Scenario Planning: bottom up, which means it is adjusted for local efficiency, because different business strategies in different jurisdictions mean collection of local metrics that matter
- Error Detection: anomalous patterns of activities, unintentional errors, and intentional misstatements

Machine-learning applications that accelerate bottom-line outcomes for the Chief Financial Officer (CFO) can encompass spend management or cost transparency. These applications provide higher visibility of:

- the priority business outcomes that must be achieved
- the capabilities required to achieve them, and
- the means to accelerate the development of those capabilities in order to realise business value faster.

Mindbridge AI, a Canadian industry leader that has also attracted investment from the Bank of England, is essentially reshaping the audit function with their AI Auditor tool. It can minimise financial loss, reduce corporate liability, and focus on providing higher value services to their clients.[402] The company has great founders focused on AI for good, and Alex Benay joined their leadership team in 2019 from his previous role leading the Government of Canada's Information Technology Strategic Plan.[403] Fluidly, a London-based company, focuses on cashflow forecasting and management in real time to help the finance function take advantage of personalised financial recommendations and decide to select the most appropriate one.[404] Prophix, a Canadian company that provides sophisticated solutions for budgeting and planning, reporting, financial consolidation, and close processes, launched an AI-powered corporate performance management in 2019, a virtual financial analyst that goes though vast amounts of data and that engages through voice and text options.[405]

Uber uses machine learning to model their real-time decision making. They claim that they have addressed some of the core challenges with annual cycles forecasting:

- possible misalignment between strategic business planning and operations
- discard real-time events
- 12-month cycle is not always a realistic and accurate basis for making decisions during the cycle

Uber's model is rolling forecasting with a range of benefits:

1. real-time adjustment to the environment
    - analyses, adjusts, and improves within the data dynamics
    - dynamic cost allocations
    - exchange information and alignment with strategic planning, operations, jurisdictions, and departments
    - Board/C-suite real-time visibility and real-time decision making
2. combines human intelligence and machine intelligence at all times
3. scalability to new jurisdictions as the business expands to new markets[406]

INFORMATION TECHNOLOGY (IT) INFRASTRUCTURE

A business process that has received little coverage has been in information technology. DiffBlue a Goldman Sachs investment, has developed AI that applies to code, allowing for a broader, more complete testing of codebase, resulting in explainability and more secure and reliable code.[407] Arago is a machine-learning technology that aims to automate all IT processes. This leads to cost reduction and a lot less frustration for the entire staff when new software updates are made and they never work. Their technology reaches an average automation level of 35 percent in the first three months of deployment.

IPcenter Virtual Engineers, an IPSoft product for infrastructure incidents and alerts, reportedly impacts 1 in 10 Fortune 100's IT operations.[408] It can eliminate human involvement on up to 70 percent of level 1 IT tasks and up to 40 percent of level 2 IT tasks.[409] This efficiency leads to reduced IT costs, flexibility, and scalability.

SCALING AI PROJECTS

One of the biggest challenges in the enterprise with AI has proven to be scaling AI projects, part of a successful AI strategy implementation as discussed in Chapter 7: Strategic Adoption of AI in Business, a simplified road map to AI adoption.

Companies like Seldon AI have built notable solutions that enable scaling AI in a realistic and timely fashion and highly recommend a thorough evaluation of the current infrastructure to ensure that it is suitable for scaling AI projects.

Grakn.AI brings a novel way to build a database, the ground work of any intelligent system that can scale. Their tool applies to every domain with complex networks of information and can be doled out on the cloud or on the premises. Seldon AI and Grakn.AI are both London-based companies and run by very competent and talented management teams.

EXAMPLES OF QUESTIONS FOR BOARDS TO ASK

	TOPIC	POSSIBLE QUESTIONS
1	AI Advisers	1.Who was our adviser for AI vendor selection? 2.Do we know who advised us on the vendor selection? If it is an internal team, how certain are we that they have the needed knowledge of AI vendors and AI technologies to ensure that we select the correct system? 3.Do our third party advisers/consultants have any conflicts of interest? For examples, do they sell us their own products, receive a sales commission from the vendors, or have any particular interest in recommending a certain vendor? 4.What's the policy the AI vendor has on access to our data and the insights derived from it? 5.In the technical evaluation, are there any considerations as to the technical longevity of the systems we are about to purchase? 6.Are we buying old technologies that are already surpassed by technological advances and which may be leave us exposed? (E.g. fake fingerprints generation has been documented in 2019, which can manipulate sensors on smart phones for fraudulent access.) 7.Have we included the relevant business teams in the AI vendor selection meetings? If not, why not?
2	AI Procurement	1.How do we ensure that the correct AI system has been chosen for the task at hand? 2.What procurement criteria have we used? Who has designed the criteria? 3.Have we covered all the technical and ethical grounds?
3	AI Vendors Data Storage	1.Where and how do they hold our data? In what jurisdiction? 2.Is the vendor regulated by the same regulator as we are?
4	Accountability	1.Who is the "parent" of the AI tool? Who has the responsibility of it? Does the Service Level Agreement reflect this detail? 2. Are we able to be innovative and be agile within this SLA?
5	Diversity	1.How diverse are the teams building the AI tool? 2.What can be done to address any potential bias issues?

KEY POINTS TO REMEMBER

- Machine learning offers many tools that can strengthen customer service, such as dynamic customer segmentation and real-time personalisation of services.
- In portfolio management AI can play an important role in operations, accelerating change in the funds industry.
- Today more than 90 percent of trading is being conducted by computer programs.
- AI in sales and marketing is all about personalisation at scale.
- AI can be used in personnel department for candidate screening, candidate engagement, re-engagement, post-offer acceptance, new hire on-boarding, career development, employee relationships and scheduling, and HR compliance and case management. Deploying AI creates privacy concerns and risks, which will only magnify in the future.
- As AI becomes more prevalent, a new legal field for AI law will likely emerge.
- Document search, data discovery, and data mining remain the main time-consuming activities in financial services' legal departments, and AI tools can help.
- In financial services, as the industry is moving to fully digital, the risk function has become complex with moving targets. There are a number of AI applications that have made risk management and compliance more efficient, as predictions have become more accurate by analysing vast volumes of unstructured data and identifying patterns and relationships in the data.
- The main areas where we find AI applications in financial services are in surveillance of conduct and market abuse. Examples of conduct breaches include rogue trading, benchmark rigging, and insider trading.
- Machine learning has been applied for over a decade in credit card fraud prevention, with transactions and responses occurring in real time.
- Some of the machine-learning uses in the function of finance include: budget optimisation, scenario planning, and error detection.

CHAPTER TEN

About the Future

When I spoke to Anthony, a dear family friend, he suggested that we meet for lunch on the floor of The Royal Exchange. This location has a special heritage that dates back to 1566 when Sir Thomas Gresham, known as the founder of English banking, established on this very location London's first purpose-built centre for stock trading. When designing the building, he took inspiration from The Bourse in Antwerp, Belgium. The Royal Exchange was the initial home of the LSE (London Stock Exchange) and in the 1980s the home of LIFFE (London International Financial Futures and Options Exchange). Today it is an upmarket eating and shopping area in London, moments away from the Bank of England building.

But I am digressing. Anthony and I agreed to meet for lunch on the floor of the Royal Exchange, rather than the first-floor restaurant. Anthony is a former trader and he has a lifelong wisdom that he is willing to share. His wisdom has meaning. He chose this location. Why? I wasn't sure.

As I was studying the menu, I heard him say, "Do you see that window up there?" Unsure about what I wanted to eat, I gladly put down the menu and looked at *that window.* "That was our regulator," Anthony continued. "One single guy, sandwiched between two huge piles of paper on his desk, chain-smoking calmly while talking to us, the two traders who had a disagreement on the trading floor. He would listen to us, make a decision, get us to shake hands and agree on his decision." That sounds easy and human, I thought to myself without interrupting the story. "A verbal agreement was our bond. *My word is my bond* is the ethos on which The City of London[411] was built on. If you broke that verbal agreement, you broke the Trust, the consequences were dire: You couldn't find a job, your career was finished," Anthony adds with a steel-like tone. Anthony has many stories like this, and I am always willing to hear them—they take me back to a measure of this profession that I want to hear more about.

The Chartered Institute of Securities and Investments (CISI), the professional body for the financial services profession that I am still an active member of, has wisely chosen the same motto *My word, my bond* to guide their work. It would be helpful if they would display it more often and make it part of their logo just as a reminder that trust, ethics, and integrity are the pillars of our profession.

After I left the corporate life, I devoted four months to study trust–how it is built and how it is destroyed and to demonstrate that building that trust with technology is possible. In my corporate career, I have seen how trust is destroyed; far too often. In 2008 we, as an industry, shattered the trust that society had bestowed on us. Whether we want to accept it or not, whether we see it or not, in 2019 this trust is still in pieces. As fintech emerges as the modern way to deliver financial services, *trust* remains the core value. There is not enough talk in our industry about how we can redefine trust, to restore trust, but we must make it a priority–especially when we start implementing AI.

HOW CAN OUR INDUSTRY REGAIN TRUST ?

Ever since its early days in Renaissance Italy, our industry's core mission has been capital distribution to support society's growth. There are many conditions for capital distribution to take place; one is that information needs to be exchanged. In the 1850s John Reuters used pigeons to send information about stocks from Paris to Brussels. In 2019, we use technology platforms. Irrespective of how we go about fulfilling our mission, one constant remains: trust.

In the age of speed, when short-termism is fuelled by quarterly reports and pressure to deliver stellar financial results, when long-term thinking is penalised as lacking ROI (return on investments), and when hitting analysts's estimates remains the only metric that this industry uses to measure their success, competing in this space means that there is always a temptation to cut corners and do away with a long-term investment in trust.

As an industry, we are a *necessity* in our customer's lives, in our society. We are service providers. We are at the service of our clients. At least we should be. They need us and so we have become comfortable and taken them for granted–we treat them as our captive audience. They are. We make money off of them. They cannot do without us. They give us their money (overdraft fees, management fees, greedy credit card interest, transaction fees, you name it–we invented all those fees), and in exchange we give them reams of impenetrable communication and slam the door in their faces when they struggle financially and need our help. For years, we forced down on them a range of products (and elegantly labeled it as cross-selling), whether our customers needed them or not; it didn't really matter (think of PPI–payment protection insurance). In this forced marriage, *trust* is absolutely essential. So what happened to us? Did we lose our moral compass? If so, can we regain it? Can we redefine *trust?* How do we change the culture to redefine trust?

THE MODERN-DAY FINANCIER

In recent years, since the rise of AI, the leadership is pressured daily with nudges to continue with "innovation and disruption" and run a data-driven business, and boards are being told that they need to deliver innovation and governance while abiding by regulatory restrictions. These are bankers, accountants, and lawyers, who have for all their lives, run a bank, an insurance firm, or an asset or wealth-management company. They are now expected to know how to run a data-driven firm, which essentially means a technology firm. And they are expected to know how to do it overnight. That's not feasible.

The modern-day financial services practitioner's career and job description is shaping up to transform almost every two years. All this time, this profession is struggling "to keep the lights on," answer those never-ending e-mails, keep up with waves of convoluted regulatory requirements and complex cyberthreats, all the while trying to wrap their heads around the next emerging technology–be it blockchain, artificial intelligence, or virtual reality–and trying to make sense of how these next-generation technologies impact their revenue models, operational models, and business models. It is hard work. These require cognitive processes that we might not be able to comprehend. And that's okay. We just need to be honest with ourselves, and admit it, and then make room for those who are able to do it, who are naturally curious and willing to learn. This becomes more critical especially at the board level. If we do not bow out gracefully, we are holding back the organisations we work for from progressing and remaining competitive.

Technology advancements pressure our profession to keep up with the change. Everything happens fast–too fast. As technology brings something new every six months, the nature of risk changes accordingly. We are pushed out of our comfort zone, and we are expected to unlearn what we have been doing for years and then learn new ways of delivering our profession. The leadership is expected to be innovative and hip, to sign off on those fat cheques for redecorating our offices with neon lights, visible pipes on the ceiling, beanbags and snooker tables, as a measure of our readiness for the next wave of disruption at a breakneck speed. We are expected to accept anomalies like "we're not aiming to reach profitability"[412] as the new normal.

Some practitioners are decidedly ignoring this profound transformation. "We have been doing it for the past 65 years. We have been successful at making money doing the same thing, why change it now?" the Chairman of a financial institution asked me at the end of a strategy day, after I went to great lengths to demonstrate with hard facts how their business model is being challenged and how they can be better off long term with changing their model and adopting AI. Right after his comment, I heard a deep sigh of annoyance from the rest of the room.

Some practitioners are committed to learning and adopting change. They leave their traditional financial services employers and join the technology movement. A mid-management London executive in his early 30s told me that six months ago he decided to take the leap of faith into AI and is now seeking his own way to learn and change his career model: "There is no business with the old business model. We need to change. Our customers want us to, but there is so much inertia at the board level so change does not happen at all. They are driving the business into the ground. The internal politics and people holding on to their seat is painful to watch. We can also be more efficient with machine-learning solutions in the back office. There is so much inefficiency in the back office that we cannot do our front-office job because of how clogged up the back office operations are. Trades are not reconciled correctly, and we still need to fill in a hand-written trade sheet in three colours, the yellow goes to compliance and we need to physically take it to them to sign, and then bring it back to the trading desk." Although he is convinced about his move, I sense that he is afraid. He has decided to jump out of his comfort zone. This is courage. Courage is identifying fear, isolating it, rationalising it, and then still acting. Courage is seeing your path despite the noise and the chaos.

In this fast lane, work life is chaotic, complex, and challenging. In chaos we are bound to forget where we are coming from and where we are going. In chaos we need to find our compass. We need to define what defines us. Trust and integrity are two values that will never go out of fashion; even in the age of AI, they define us as an industry.

My conviction is that with emerging technologies we have the opportunity to redefine trust, to personalise our services at scale, and to give our clients what they need instead of what we need, while still remaining profitable. In doing so, we would transform our industry from a plain necessity to a helpful industry that shares human values like empathy, compassion, and the anticipation of needs that are the real foundation of trust in humans.

The future is decidedly with AI. Trust with technology is within our reach, and there is a business case for that.

In the aftermath of the Second World War, Siegmund Warburg, the prominent English banker, a refugee from Nazi Germany, "blended financial innovation, creativity and morality."[413] Fifty years on, these values remain timely. Niall Ferguson's superbly written and documented biography of this remarkable banker, *High Financier,* concludes in the book's introduction that the 2008 financial crisis can be "blamed on the conscious abandonment by a new generation of bankers of Siegmund Warburg's ideal–financial services based on the primacy of the client relationship rather than speculative transactions. In the eyes of the public, City of London's reputation has sunk low indeed. If ever there was a time to learn from a true high financier, this is surely it."[414]

PRINCIPLES OVER PROFITS

Our mission remains the same (capital distribution to benefit our society and make money in the process), but our purpose is different, as technology is permeating the fabric of our society and of our lives. Our purpose should be to actively direct the capital to solve some of society's key concerns like environmental risk, universal basic income, and healthcare, as well as prosperity distribution as the data-driven economy is creating new wealth and it is expected to exacerbate inequality.

Our purpose has always been to make money. It still is. Technology is a useful tool to help build a sustainable future for our society and our industry. For instance, there is no long-term value to managing a pension fund and using it to invest in fossil fuel firms without carbon offsetting, if we do not make sure the planet is liveable, if the window in which we can save the planet from environmental destruction has already closed. Similarly, our entire business model will be affected if inequality grows even deeper–people will not be able to save into their retirement. There will not be a pension fund to manage. We will not be making money. I think we have moved past "paying lip service to sustainability," and now we can understand that linking the financial system with sustainable development creates sustainable banking, a new business and value generation model.

"The business of business is business," professed Milton Friedman in the 1980s and every business school echoed his opinion. I remember that it never sat well with me when I received my business degree. Next, in the early 2000s Corporate Social Responsibility (CSR) classes were introduced in business schools. Recently, this narrative has gained a new meaning. That is ESG+T[415] that is Environmental, Social, Governance + Technology. This is the new definition of good governance, and it is permeating boards and leadership teams.

Like a tsunami, the 2000s financial crisis laid bare a range of structural issues and inequities in our society like income inequality and the insurmountable gulf between the uber-rich and the vast majority of the workforce. This realisation of the many of their futile effort to cross that gulf eroded the trust in the capitalist way, so much so that in 2016, a Harvard study highlighted that 51 percent of Americans aged between 19 and 29 openly do not support capitalism. Internet 2.0 came to fruition after 2008, bringing the automation of jobs and making the working class feel ever more endangered, thus compounding the structural issues of poverty and populism. In 2008, Bill Gates gave a talk in Davos about "creative capitalism." In his speech at the World Economic Forum, he articulated that "the genius of capitalism" lies in its ability to "[harness] self-interest in helpful and sustainable ways. We need a system that [...] would have a twin mission: making profits and also improving lives for those who do not fully benefit from market forces." In December 2016, after the US election, and at the encouragement of Pope Francis, a group of 100 CEOs met in Rome and spent a day deciding on how the private sector could maximise social impact. In August 2019, The Business Roundtable (BRT), an association of the chief executive officers of nearly 200 of America's most prominent companies, announced their commitment to a new purpose, creating "value for customers," "investing in employees," fostering "diversity and inclusion," "dealing fairly and ethically with suppliers," "supporting the communities in which we work," and "protecting the environment."

The Task Force on Climate-related Financial Disclosures (TCFD), led by billionaire Michael Bloomberg, aims to raise the level of transparency reporting on climate-related information, to enable investors and lenders to make more ESG-informed decisions. In 2017, 237 companies, with a combined market value of more than $6.3 trillion, publicly committed to these TCFD recommendations. The European Commission agenda supports this initiative and put forward a strategy to align its financial system with the EU's climate and sustainable development agenda. A number of banks have also committed to sustainable banking: Westapac (Australia), KiwiBank (New Zealand), Bank Sarasin (Switzerland), Triodos Bank (Netherlands), and Bank of America (United States).[416]

AI HYPE? REMAIN VIGILANT

AI has been around for more than 60 years. Only in recent years has it reached out beyond the confines of academic life and research laboratories into our society, government, and businesses. What we currently have access to is *Narrow AI*, narrow in that it is applied to one process as opposed to *General AI* otherwise referred to as *Artificial General Intelligence (AGI)* which works seamlessly across a range of unconnected processes and fields of knowledge.

The breakthrough advancements in deep learning and its applications have created a hype and a distorted perception of what these technologies are and what they can deliver. "This wave of AI hype will have dire consequences. Not just for our field [NB: AI field]; for public safety. Overselling the progress and capabilities of AI leads governments and companies to adopt shoddy 'AI' solutions (which contractors are more than happy to sell). And blindly trust their predictions,"[417] warns Francois Chollet, a Google deep-learning scientist and the creator of Keras, a neural networks library. A host of ethical implications of ethically challenged AI systems have already surfaced, magnifying the historical bias and inequality without any accountability of their outcome. There are many voices in the AI research community pointing out the flaws that current AI technologies have. They are further limited as they lack context understanding and some have their outcome predefined. Deep learning can find patters in data but it cannot explain how they connect. In the Book of Why by Judea Pearl discusses the casual limitations in deep learning and it is a book that I highly recommend. There is a hidden technical debt in machine-learning systems amongst which configuration issues and hidden feedback loops cause some of the issues identified.[418] We also have a control problem for this technology. For the above reasons, Chapter 4 discusses technology maturity cycles.

This book highlights the main concerns and why our industry needs to remain vigilant and observe how AI challenges other sectors. Boards and management need to keep asking the hard questions and ensure that they can see beyond what AI vendors want to sell them.

This book also lists a wide range of vendors delivering solutions in our industry. Not all are truly AI companies, although most of them claim to be, but they have proven useful in the business context they are used and this was a good-enough filter to select them. If these vendors persist and are well managed companies, over time they may build quality datasets, improve their technology, and become true AI companies. Some vendors describe themselves as AI powered or AI driven. Some other vendors do not even want to describe themselves as AI and instead use Augmented Intelligence, perhaps a reflection of the historical disagreement on the definition of AI.

However they describe themselves, our industry is well advised to always be vigilant and to ask robust questions at the procurement stage: technical, ethical, and business questions. Our industry needs to insure that what we bring in or develop in-house AI solutions that are robustly rooted in ethical AI principles and that use the correct type of algorithms and datasets to support the organisation's business growth plans. We can build profitable organisations with ethical AI, especially in our industry even with the current AI technologies. If we open the door of our industry to wrongly designed AI systems, we will likely lose the trust of our customers and damage our businesses. Our industry "needs to know what we say yes to" to paraphrase Kay Firth-Butterfield of the World Economic Forum.

PRIVACY, IDENTITY, DATA SHARING, AND DATA PROTECTION

There will continue to be an increased tension between delivering AI projects and privacy. There is no proven and clear path to solve it. Personal data has become a core source for value and innovation, as it is monetised by a number of personal data brokers. There is a generally accepted theory that the more data the more value to be created. This assumption is yet to be proven, as regulatory requirements and technological challenges are weaved in tightly and hard to predict. *Trust: Data: A New Framework for Identity and Data Sharing* edited by T. Hardjono, D. Shrier, and A. Pentland is recommended reading for decision makers in our industry to better understand how academia and practitioners see how emerging technologies might be able to address what Professor Pentland refers to as Society's Nervous System data, personal privacy, and information security.

In 2011, the World Economic Forum published a report titled *Personal Data: The Emergence of a New Asset Class*, which I also recommend reading. Consent and data protection are areas that continue to raise fundamental legal, privacy and ethical questions, and they need to be well understood in the context of each organisation and managed as a business risk. The same questions are raised by the UK's companies register, a freely accessible database of companies directors, name, date of birth and companies' account. This puts a target on companies directors, in the name of transparency, without even providing directors with the visibility of who is looking them and their businesses up.

DATA VALUE STRATEGY

Boards need to be able to understand what value their organisations can derive from the data they hold, they own, and they have access to. Identifying a key range of attributes would enable a better understanding of data value and this would impact their business strategy and revenue model:

- existent data: audit of (1) decision-making processes like pricing, cybersecurity, planning, operations, and business development, (2) existent data points used, (3) people who rely on data analytics for decision making, and (4) support people who prepare data and draw insights
- real-time data: where it is needed (e.g. financial forecasting) and where else it can add value
- unique data: why we have it and who wants it (internal and external)
- essential data: data needed in a specific process optimisation
- prized data: data that the business depends on and is most at risk in a cyber attack
- regulated data: data that falls under the incidence of regulatory requirements and that needs to be managed under an auditable protocol for collection, analysis, compliance, and strategic decisions (e.g. pricing strategy using behaviour prediction)

BLOCKCHAIN, GDPR, AND DATA SUPPLY

Blockchain technology has promised a lot and in some cases, has delivered. Every large organisation in financial services that I know, has blockchain or Distributed Ledger Technology (DLT) projects. Chris Skinner eloquently stated that in financial services, blockchain has been a solution looking for a problem. In data management there are a few projects worth mentioning, including:

(1) GDPR: The European Union's General Data Protection Regulation (GDPR) marks the most important change in data privacy regulation in over 20 years. It has a ripple effect on businesses offering digital services to European clients. Consumers with bank accounts in Europe fall directly under its influence. BigChainDB is a blockchain database where users insert data that is encrypted and then tokenised. The user can then send a token that can be used to access this data. Data can be stored encrypted on a separate decentralised or centralised file system. BigChainDB helps with satisfying GDPR requirements of increased security, transparency, and resilience based on the use of immutable, distributed records. For further reading, I'd recommend the article "Blockchains and Data Protection in the European Union" by M. Finck of Max Plank Institute.

(2) Distributed Datasets: Significant data is stored in silos within organisations and within the industry. Instead of breaking these silos down to join the data in one central repository, blockchain technologies can be used to create an infrastructure for distributed datasets that are used to train AI models. This is also one of the promises BigChain DB offers.

(3) Decentralised Data Exchanges: It is worth noting Ocean Protocol, a project under the BigChainDB name,[419] that was designed to democratise access to data, to enable AI developers to access the data they need. This ultimately implies democratising AI and breaking the monopoly on data held by the big technology firms. If you consider data an asset class, Ocean Protocol aims to be to data what the London Stock Exchange is to shares and bonds–a trading place where data is bought and sold, with data sharing in a trustful, traceable, and encrypted way.

CYBERSECURITY

Cyber attacks have become more sophisticated. Bots (algorithms) can be programmed to perform cyber attacks, to attack and defend, to repair the protection wall in software. Significant funding has gone into cybersecurity. Despite these developments, cyber criminals continue to perform successful attacks. Some are innovative, but most merely exploit the software's common programming errors or humans' naivety and carelessness. Cybersecurity training should be regarded like health and safety, a compulsory ongoing training to include the latest type of attacks. For instance, GANs generating fingerprints that can fool smart-phone sensors and teach people how to be alert and vigilant. Currently, there is no consistent cyber hygiene across organisations. People remain the first weak point to break in. We need to change how to protect ourselves from cyber crime. It starts with us.

AFFECTIVE COMPUTING

This is an emerging field in AI. It aims to measure emotions in humans through evaluating changes in facial expressions, modulation of voice, body language, and gestures. These are important abilities in conversational AI used in customer care and personal assistants. The emotional meaning is conveyed 10 percent through words, 35 to 40 percent through voice, and the remaining 50 percent through body language. MusimapAI profiles people's emotions depending on their music taste. Emoshape built the first emotion synthesis chip that enables machine awareness as a first, including pain, pleasure, frustration, and satisfaction, in addition to Paul Ekman's psycho-evolutionary theory that identifies 12 primary emotions.[420] Being More Digital builds predictive models to support insights to customer loyalty and satisfaction, using real-time videos of human emotions and movement.[421] Affectiva, a leader in affective computing, collected over six million faces from 75 countries, and nearly two billion factual frames and use machine learning to analyse them. One of their earlier applications is in partnership with Brain Power, using Google Glass to assist autistic children in identifying and correctly understanding human emotions. Other commercial applications are evident in customer care and customer engagement.

5G CELLULAR NETWORKS

While writing this book, there was a whirlwind of discussion around what 5G means and how it will transform internet connectivity with speed and a range of other ICT industry requirements. 5G networks might prove insufficient if they lack AI functionalities, with the aim to build intelligent networks.[422] "AI and 5G are heralds for the coming data age," Steve Koenig stated at the CES 2018. 5G is expected to bring improved speed, reduce latency, and realistically provide unlimited data plans. This speed and unlimited data package will enable IOT, augmented reality, and smart cities[423] technology, which are bandwidth hungry. It will also improve the speed at which we are transferring information and enable complex algorithms to work faster without latency. With a 4G wireless network, we can download a two-hour movie in six minutes. With 5G, we would be able to do that in less than five seconds. How does this translate in our industry? It may be that hologram financial advisers can become a "reality" in their clients' homes, saving them the travel.

INTERNAL ENGINEERING CAPABILITIES

When the Government of Singapore brought in Chief Digital Technology Officer (GDTO) Cheow Hoe Chan from the banking sector in 2014, his mission was to build meaningful independence from technology vendors that the government was heavily reliant on and to build an internal capability to support in-house R&D and deployment of projects. "Internal engineering capability allows the island nation to flex its muscle in negotiations,"[424] he quipped when asked why do it. There is value in this approach, and it is a lesson to learn from for our industry decision makers. Building internal engineering capability means building value into your business and independence from vendors. The most valuable companies in financial services will be those that will eventually become full-stack companies, vertically integrated technology to support their own ecosystem.

In 2019, when it came to public attention that 106 million of Capital One's clients had their personal data hacked, many raised the question as to what extent AWS (Amazon Cloud Service) was responsible for this. The bigger questions may be:

- To what extent can our industry entrust client data to third-party cloud vendors when they are not subjected to financial services regulations?
- How vulnerable is our industry as we migrate our data, our business information, and our clients' data to third-party cloud vendors?
- Is now the time to have an industry-wide initiative and build a platform to host our data as an industry (clearly segregated access), while sharing the cost of setting up and maintaining it?

Finally, when a credit-card supplier like Capital One is hacked and the personal data of 106 million people is compromised, when their spending history may have also been hacked (who really knows), one has to wonder who needed this data and what they can do with it. Profiling people's behaviour to influence and manipulate their choices is more potent with financial data than with social media data (i.e. the number of likes of photos of cats, dogs, babies, and holiday pictures).

DATED INFRASTRUCTURE AND PATCHWORK UPGRADES

I have lost count how many times I have heard CIOs, CDOs, or CTOs expressing concern about how dated infrastructure will not be able to support more layers of AI systems. They have been patching up this infrastructure, but patchwork can only take you so far, plus it is usually very expensive in the long term.

A board director concluded to me recently that it is easier to build a digital bank outside the confines of a dated infrastructure "it just does not make sense to do it." It may be even cheaper, too. The number of billions the banks are committing to digital transformation in 2019 is impressive. Santander just announced $22.5 billion for digital transformation[425] and Commonwealth Bank in Australia committed to spend $5 billion. JP Morgan in 2018 set aside $10 billion for digital transformation, out of which $5 billion is solely for AI projects. So, what is this money actually doing? It's one thing to spend it, but an entirely different prospect to achieve long-term growth for your organisation. In a private setting, one of the top digital consultancies said that a mere 25 percent of digital projects are successful–in other words 75 percent of the money spent on digital transformation is a waste of resources.

This begs an honest conversation at the board level: Why is this happening? Do we know what we are spending all this money on? Do we actually know what we are doing? Is all this patchwork worth it?

BIOENGINEERING: YOUR FUTURE CLIENTS ?

A field that has yet to catch the media's attention is bioengineering, a discipline that applies engineering principles of design with biomedical technologies. Streaming our thoughts through neural interfaces that will enable humans to communicate to other humans and machines are expected to revolutionise humanity. This poses a wide range of technical questions around bio-compatible hardware. One of Elon Musk's breakthrough endeavours, Neuralink, is aiming to successfully connect humans and computers with ultra-high bandwidth brain-machine interfaces, in the shape of chips that are installed under the skin. They announced that in 2020 they hope to start the technology on humans. Their immediate applications are in the medical field to help patients with brain and spinal cord injuries or congenital defects.[426] The longer-term question is "what does this mean to our industry?" Just for the sake of argument, assume for a moment that a client gets these implants, gains super-human abilities, and at some point might even be able to run his or her portfolios better than a discretionary wealth manager? What would be the impact on our workforce? On our business models?

QUANTUM COMPUTING

Earlier in the book, I addressed the main computing approaches. They are all aiming to achieve High Performance Computing, which is the current focus of the semiconductor market. At the moment, there's still insufficient computational power available to address some of the most complex problems but progress is being made. A new type of computing is necessary. Quantum computing is hailed to provide the computing power that is needed, to shorten time-to-solutions from months to seconds, and in time, if the currently high costs of design and development come down, it will likely be widely adopted.

The implications are fundamental, and it will require redefining how software is built, how data will be stored for access and processing (currently quantum computers cannot store data), how encryption will work, how development of AI systems will emerge, and how pricing models will be changed. The current computers store and process data using bits. Bits have only two states: 1 or 0. As an analogy, they work in red state **or** green state. Quantum computers store and process data using *qubits,* which have a combination of these states. In practical terms, useful applications will emerge when quantum models have minimum 100 *qubits,* while at the moment 49 was the highest reported number of *qubits* achieved by Google.[427] To keep the analogy, depending on how much red you mix in, you'll get a different state each time, and this is where the computing power progress is achieved. An easy-to-follow, in-depth explanation is provided by IBM Research, which I recommend to those who wish to further their knowledge: *www.research.ibm.com/ibm-q/learn/what-is-quantum-computing/*.

Some specialists argue that 100 *qubits* can be achieved in 20 years' time. Whether it happens in 20, 30, or 50 years, it is not certain. What is certain is that IBM, Microsoft, Google, and Cambridge Quantum Computing are currently investing substantial resources to develop these capabilities with notable progress in their outcome.

What does quantum computing mean to our industry? Everything. It would open a host of new opportunities to deploy AI, work with the data, and deliver long-term business growth. Boards are well advised to include a scenario analysis of quantum computing applied to their business models and growth projections. I spoke to Roberto Desimore of BAE Systems about the progress made in this space. In 2018, while on sabbatical from BAE, he led a groundbreaking feasibility study on "existing AI planning and scheduling techniques that could be enhanced by quantum algorithms to deliver optimised plans and schedules in real-time for complex tasks, that are considered too demanding for conventional classical computers." The project titled "Quantum algorithms for optimised planning/scheduling applications" was sponsored by Innovate UK and supported by the UK National Programme for Quantum Technology and I recommend it for further reading.

NEXT GENERATION LEADERSHIP

Chapter 1 is the only co-authored chapter in this book. Why? My co-author, Alexander J. Deak, is a 21-year-old student in his final year at The University of Oxford. His career of choice is corporate finance, although for some recruiters, the fact that he is studying theology, logic, ethics, and philosophy of science makes it appear as if he has gone astray from his objective. He didn't go astray. In fact, I believe he is the type of leader we will be seeing in financial services in 10 years. The financial services industry has wrongly favoured recruiting juniors who have an economics background. This is fundamentally flawed.

This also shows how little recruiters actually understand about how AI is redefining our industry. And this is why recruitment is driven by short-termism and a lack of understanding of what leadership in an AI-driven organisation needs. Currently, organisations are already encountering a widening gap at the intersection of the three defining strategies (1) business thinking, (2) data-engineering thinking, and (3) human-values thinking. This gap is rapidly widening. The symptoms will be adoption of technology that does not speak to clients, which means wasted resources. Our industry vocationally trains all of its entry-level juniors, usually across two years, yet there is no equivalent training in humanities–philosophy and theology.

As an economist, I would argue that it is easier to teach a theologian finance than to teach a financier theology embedded in positive human values. Our industry has proven to lack discernment in many areas to do with its impact on humans. The more connected we have become with social media and instant messaging, the more disconnected we have actually become as a society. One of the reasons for our disconnect is the lack of *empathy* and *genuine understanding* of others' individuality, religious beliefs, and human aspirations.

In many organisations, their leadership also seems disconnected from the business, its employees, and its customers. As our industry becomes AI-driven, we need to ensure that we use technology to bring us closer to our customers and to bind our staff closer together and to the organisation; the new leadership type must have the technology, business, and humanist values embedded in each leader. Outsourcing this knowledge will not be enough. This is essential, so that we do not build technology (and sign off on billions of expenditure) that does not connect with humans, our customers, and our employees.

CUSTOMERS' EXPECTATIONS: THE MONTHLY DATA STATEMENT

This expectation is not determined by age group. It is determined by customers' understanding of the value of their own data. I predict that customers will expect to receive a *Data Statement* from their banks similar to the usual monthly account/credit card/ mortgage/savings statement.

Customers will want to know for instance (1) what data financial institutions hold on them and where is it stored, (2) how they use it (including when it is anonymised) and for what purpose, (3) how to withdraw consent to use it, (4) how data privacy is withheld, and perhaps (5) how banks are using customer data to provide improved services. Banks are opaque about their usage of customer data. They do not know how to open up and produce such a *Data Statement*. In this context, it is worth mentioning that Intel is making notable progress in transfer learning (approach to machine learning); coupled with their 2019 breakthrough in data-centric computing, Intel technologies will be able to addresses many data-privacy issues. I would also argue that a Data Statement is the by-product of a larger conversation on the future of identity and the role the financial sector should play as custodian of their customers' data.

For instance, when Mastercard agreed to provide Google with the credit-card data of millions of customers for Google to enrich their already vast datasets for advanced targeted advertising, we, the customers, need to know–there is a new level of targeting upon us that uses our personal data to generate revenues for other people. It can be downright frightening when you search for a product and then adverts for that very product category follow you everywhere you go like a stalker. With financial data from our transactions, such advertising is pushing the boundaries to a place where many people find it exploitative, dishonest, and intrusive. Putting aside the strategic question "Do we want our customer subjected to this?", the least banks can do is to let us know in a monthly data statement. Customer data aggregations, tracking, and reporting might look like a new business model. There are executives who believe that a *Data Statement* is a fair expectation.

AI SUPERHUBS

In her bestseller, *Superhubs,* Sandra Navidi illuminates how the financial elite's exclusive networks of influence, formed by bank CEOs, fund managers, and billionaire financiers, have been built, exist, and rule financial services ecosystems and our world. In a similar way, I would like to highlight that there is a clear trend to replicate the same strong networks formed by the congregation of similarly influential people, only this time to decide how the future with AI will look. They are creating what I call the AI Superhubs either in academia, research, or policy making. The AI Superhubs manifest themselves as exclusive clubs that organise regular invitation-only influential events where trends are formulated and agreed. The invitation to attend is subject to, in some cases, paying exorbitant membership fees.

Ethical AI, a technology which can bring about trust, is not going to be shaped by an exclusive handful of people or organisations who have the *white badge,* the symbol of a substantial fee they have spent on an exclusive membership. It might be rather pleasant to mingle in the same exclusive circles of people connected only to the stratosphere. However, ethical AI is also about diversity, inclusivity, removing bias, and impartiality. It also means coming down from the stratosphere down to planet Earth where this technology affects normal people. Normal people, the vast majority on this planet, have a host of problems that are vastly different from those of the privileged. The membership to these AI hubs should not be about *the white badge*. It should only be about one single badge: the badge of honour, integrity, professionalism, and ethics of people from all races, business backgrounds, job seniority, or education, people who are interested in shaping the future with AI. If these exclusive AI gatherings do not open their doors to be inclusive when deciding not just consulting on how to influence policy making, then they will unknowingly introduce bias and more likely promote a distorted agenda, that produce anomalies that will be difficult to correct in the future. I implore these AI Superhubs to gather more financial sector practitioners of all level of seniority and ideally impartiality, and to reconsider being more inclusive.

REGULATING AI

Brad Smith is the President and Chief Legal Officer of Microsoft. He is also the co-author of the foreword to the *The Future Computed–Artificial Intelligence and Its Role in Society,*[428] a book published by Microsoft, a book where he professes that "AI technology needs to continue to develop and mature before rules can be crafted to govern it. A consensus then needs to be reached about societal principles and values to govern AI development and use, followed by best practices to live up to them. Then we're likely to be in a better position for governments to create legal and regulatory rules for everyone to follow." This line of thinking may be applicable to AI technologies in other sectors, although I doubt it is. However, in financial services we need to consider a different approach. I believe that, as an industry, we need to define our own future with AI (we have seen enough of what it can achieve–for instance, think facial recognition and the issues it has created or racial discrimination). Many board directors think that an AI crisis is predictable and that will become the catalyst to AI governance in our industry. Why wait when you can influence the rules of the game early?

It is imperative that we put guiding principles and law in place right now and implement them. If we wait for the AI "technology to evolve," meaning that technology firms do whatever they see fit without a regulatory framework–and only then call in the regulators to regulate them–it will be too late for society to regulate and shape technology to support humanity rather than technology firms' profits. For instance, in 2019 Facebook is asking the regulators to regulate them, after Facebook's influence created so much irreversible damage. It has already proven difficult to regulate Facebook or Google's adverts promoting financial products targeting savers[429] or Facebook's invasion of privacy, use of personal data, or Amazon's facial recognition tools.

As an industry, we also need to define our future with AI. AI technologies are powerful because they are scalable in a short period of time. Experts warn that AI evolves 10x every six months, while data grows at the same rate every five years. The pace of change is unprecedented. AI technologies will determine us if we do not determine them first. As I keep repeating, "Algorithms have parents." We are responsible for how they develop. Our industry needs to put clear regulations in place to avoid predictable abuses of personal data, algorithm design, and trust.

While AI is responsible for creating a new order in financial services, the rules remain the same in our global village: trust, integrity, and putting the customer first.

Plus ça change, plus c'est la même chose.

APPENDIX

Developments in AI:[430] A Brief History of Achievements

1637
Rene Descartes, the French mathematician and philosopher, stated that it will never be possible to build a machine that thinks the way humans do. According to him, machines think through their electric circuits and components not through their thought.

1840
Ada Lovelace was the first to recognise the full potential of a computing machine, and she was the first to publish an algorithm to be used on such a machine. She is one of the first computer programmers. Charles Babbage proposed a mechanical general purpose computer.

1943
Warren McCulloch, a medical doctor and psychologist, together with Walter Pitts, then at the University of Chicago, recognised the possibility of modelling the brain mathematically.[431] This paper is considered the first step in building Artificial Neural Networks (ANNs).

1949
"Organisation of Behaviour" was a seminal paper that aimed at creating a taxonomy of the biological brain and human learning, and for the first time, it posited that memory is built through repetitive use of neurons' synapses. This insight in human learning opened the possibility for computer scientists to understand how best to try to train Artificial Neural Networks (ANNs) by emulating the human learning process.

1950
Alan Turing wrote a paper that talked about the Machine and Turing Test, and he posed the question, "Can machines think?"

1950s
Norbert Wiener's work was one of the first to theorise the link between human intelligent behaviour, which he identified as being the result of feedback mechanisms, and machines that can simulate these mechanisms.

1955
The Logic Theorist was the first attempt to show the link theorised by Wiener.

1954
First tests on artificial neural networks (ANNs) formed of up to 128 neurons achieved accurate outcomes in recognising patterns. This development built on the earlier work in the University of Chicago by McCulloch and Pitts.

1956
John McCarthy organised the Dartmouth Conference, building on 20 years of work, and congregated a large number of computer scientists to discuss the progress and direction of how machines can simulate "learning or any other feature of intelligence." This conferences serves as the official birth place of AI, as McCarthy coined the term "artificial intelligence."

1958
Frank Rosenblatt of the University of Cornell introduced the perceptron, an early form of neural network, which provided the foundation for the next state of developments in AI–expert systems–with the input from Stanford University professor, Edward Feigenbaum.

1958
Early genetic algorithms experiments.

1964
Natural language processing emerges as a field within AI; machines understand natural language minimally enough to solve word problems.

1966 - 1972
The Stanford Research Institute (otherwise known as the Artificial Intelligence Centre at SRI International) started their research on the mobile robot called Shakey, which could perform tasks that required planning, route finding, and rearranging of simple objects. Shakey is now on display at the Computer History Museum in Mountain View, California.

1969
Marvin Minsky and Seymour Papert, both professors at MIT. wrote the book *Perceptrons: An Introduction to Computational Geometry*. Their book set the tone for the next generation of advancements in artificial intelligence. They also raised the discussion on how little science is known on how the biological brain works.

1973
British government ends funding for AI research, based on Lighthill report's findings.[432]

1974-1980
The First AI winter–funding dried up, and no advancements were made.

1980s
A new wave of investments mark an AI revival. USA, Japan, and the UK competed heavily in AI funding. DARPA spent over $1 billion on its Strategic Computing Initiative. Japan invested $400 million through the Japanese Fifth Generation Computer Project. The UK invested £350 million in the Alvey Program. The first steps were taken in quantum computing.

1980
Expert systems or knowledge systems emerged as a new field within AI.

1982
James Simons founded the quantitative investment firm Renaissance Technologies, and this marked the first milestone into the adoption of AI in financial services.

1987
Chase Lincoln First Bank introduced an expert system (PFPS) to provide objective, affordable, expert financial advice to individuals with household incomes ranging from $25,000 to $150,000, and up. It covered investment planning; debt planning; retirement savings and settlement of retirement plans; education and other children's goal funding; life insurance planning; disability insurance planning; budget recommendations; income tax planning; and savings for achievement of miscellaneous major financial goals.[433]

1987 - 1993
Second AI winter when the funding for AI research dried up.

1988
DE Shaw has an early adoption of AI in hedge funds investment management.

1990s
Lisp and Prolog were the main languages developed as key to Artificial Intelligence.[434] The AI community renewed its interest in neural networks.

1990

- Brown, Nielson, and Phillips introduced an overview of integrated personal financial planning expert systems (knowledge systems). PlanPower provided tailored financial plans to individuals with incomes over $75,000.[435]
- Neural-networks-powered devices read bank cheques and recognised the amount to be paid.

1996
The Financial Crimes Enforcement Network (FinCEN) monitored money laundering, using its own AI system.

1997

- Long Short-Term Memory (LSTM) was described in a research paper by Schmidhuber and Hochreither. LSTM was proposed as a method to boost ANNs by adding a memory function. The researchers proposed a circular improvement that would use the memory of previously learned activity. Naturally, this improvement augments ANNs with each added memory.
- IBM vs Kasparov. Deep Blue defeats Kasparov in a set of chess matches. IBM's stock price rose to unprecedented levels.
- Winton Capital, the hedge fund, was founded. They specialise in momentum trading.

2004, 2005, and 2007

Autonomous vehicle technologies were formulated and tested during three yearly contests organised by the US Government's Defense Advanced Research Projects Agency (DARPA) in the Mojave Desert, as well as urban environments on a military base. The technology tested was LIDAR (Light/Laser Detection and Ranging), a sensing technology that was primarily used for military purpose of mapping the environment and responding to it. LIDAR is essential in autonomous (self-driving) vehicles. Ten years on, in 2017, LIDAR technology became the object of a heated legal dispute between Uber and Google over proprietary technology that one of Google's former employees was accused of taking when he left Google.

2006

- The year when artificial neural networks (ANNs) were rebranded as *Deep Learning*. This coincided with the groundbreaking work by Geoff Hinton, who introduced an algorithm that was able to fine-tune the learning of ANNs with multiple layers or deep layers.
- Important development year for image recognition. ImageNet was set up at the University of Stanford. The project was initiated by Fei-Fei Li, a professor at Stanford. ImageNet was a database of one million images and it is what made the training of deep-learning models possible. This is a landmark moment because it provided deep learning with what it needed (troves of data–in this case images).

2009

Google's first self-driving car

2010

S&P crashed 8 percent, before a rebound, during a 36-minute window. This was a "flash crash," a United States trillion-dollar stock market crash, which started at 2:32 p.m. on 6th May 2010. A flash crash is exacerbated as computer trading programs react to aberrations in the market, such as heavy selling in one or many securities, and automatically begin selling large volumes at an incredibly rapid pace to avoid losses. Flash crashes can trigger circuit breakers at major stock exchanges like the NYSE, which halt trading until buy-and-sell orders can be matched up evenly and trading can resume in an orderly fashion.[436]

2011

IBM's Watson wins against two champions in the TV show *Jeopardy*, by using a database of 200 million facts and figures. IBM collected the winner's prize of $1 million.

2012

- Image recognition technology marked a landmark development when an AI algorithm trained on thousands of screening images helped a team of pathologists to make more accurate diagnostics of breast cancers.
- Knight Capital deployed unverified trading software and lost $440 million within 45 minutes.

2013

Image recognition technology continued with more successful advancements with the work of Vicarious, an AI start-up, which used a recursive cortical network, a type of neural network, to crack CAPTCHA (Completely Automated Public Turing test to tell Computers and Humans Apart).

2014

- Vital was the first-ever board member algorithm appointed to a board of directors, to help with investment decisions. The sheer amount of data, and the insights it was able to provide, were considered of essence to the board. It was also expected to vote on investments.
- Reinforcement learning paper from Google refined this technique to a new dimension to machine learning using reward.
- Generative Neural Networks was an approach to unsupervised learning discussed in a paper by Ian Goodfellow.
- MAN group started managing client money using its own AI.

2015

- Natural Language Processing (NLP) advancements, with Google reporting that it used Long Short-Term Memory (LSTM) to reduce the error rate in language understanding by 50 percent. NLP/language understanding is used in Alexa or Cortana.
- Baidu, Inc., a Chinese multinational technology company, recognises two languages.

2016

- March - Google AlphaGO wrote history
- April - Alibaba's AI
- Human parity - object recognition from Microsoft

2017

- DeepStack wins at poker game
- Human parity - speech recognition from Microsoft
- Two Sigma, the hedge fund, used machine learning and had $5 million under management.
- China announced they want to lead the world in AI by 2030.

2018

- Human parity - machine translation MS
- Human parity - speech processing MS
- August - DARPA announced MediaForensic, the first forensic tool created with AI, developed to identify fake news, fake videos, and AI-generated forgery (Knight, 2018).[437]
- Google launched Duplex, the voice assistant that mimics to perfection a human's ability to converse. It would pass the Turing test.
- MiFID II took effect.
- GDPR took effect on 25 May.
- Ant Financial, the AI company, announced that they have 850 million active retail banking clients and 11 million in small business accounts.
- Fujitsu announced that they have developed an explainable AI for credit assessment for small business loans assessments.
- UK announced country-wide strategy and exact plans for implementation of AI and ethical AI.
- Baidu was the first Chinese company to join the Partnership for AI, a US-led consortium established to devise safeguards for the development of artificial intelligence. Baidu was the first Chinese company to heavily invest in AI. It was shortly followed by Tencent and Alibaba.

2019

- British government launched investments in PhD and Masters programs in AI.
- US announced their AI strategy.

ACKNOWLEDGEMENTS

Writing a book is a solitary act, yet it takes a large network of people to publish it. While this book was mostly written during my business travels–in airport lounges, on airplanes, and in hotel rooms–I have always had a support system of people who have helped me along the way. To them, I would like to dedicate some of the chapters in this book.

Chapters 2, 3, and 5 to Hertford College, where I researched core AI concepts, read computer science books, and spoke to computer science professors who helped me clarify many technology questions. Alice Roques generously opened the library for me and offered me unconditional support. Steve New's encouragement gave me the impetus to dive into this project with courage. Chapter 7 was finished during a flight from London to Kuala Lumpur and I couldn't be more grateful to João and Qatar Airways for their superb service and giving me the comfort to focus on writing.

Chapters 8 and 9 were finished during my stay at The Banjaran in Ipoh, Malaysia. This place is a true gem in the jungle. I'd like to dedicate these chapters about how AI is transforming our industry to Tan Sri Dato' Seri Dr. Jeffrey Cheah, the founder of Sunway Group and of The Banjaran Hotel, whose work, career, and achievements were a subtle inspiration during my stay. Seeing his vision of Ipoh as a smart city with a research centre and sustainable infrastructure made me understand why his work is so valued in the Ipoh community and beyond. Raymond Koh Fei Ming and his superb team looked after me impeccably and that added that extra spark of inspiration to write. Terima kasih banyak banyak! Thank you so much!

I am thankful to the talented people who took my manuscript and expertly turned it into a book. I am deeply grateful to my editor, Lara Asher, for making impenetrable technical writing easy to read in layman's terms, for her confident grasp of a complex project and its content, and for guiding me where more clarity was needed. Her vast editorial experience and competence was essential to get the manuscript ready for printing. I couldn't have asked for a better partner to start and finish editing this book. I'm grateful to Ardi Kolah for his valuable feedback on aspects of AI risk management in the enterprise, especially data privacy, as well his practical advice on how to bring this book to life. The book's first reviewers, John Meinhold, Colin Bennett, Philip Courtenay, and Alexander J. Deak provided valuable editorial feedback. The next edition of the book will include more reviews, which we haven't been able to include due to the printing deadline.

The book that you are currently holding, beautifully designed, printed and bound, exists thanks to Blissetts. Gary Blissett was infinitely supportive while I was finishing the manuscript and Robbie Patterson who designed it all with an expert eye for detail and aesthetics. The Royal Warrant received from The Queen is a testimony of the exquisite work that Blissetts does. I couldn't have found a better home for my book. Nigel d'Auvergne expertly indexed the book, and his years of experience were of essence to complete this task accurately and on time. Emily Birchenough approached the manuscript with a fresh pair of eyes for the final proofreading checks.

There are a few people who have been pivotal in my journey, which led me to writing this book: Lord Dobbs, the author of many best sellers, amongst them *House of Cards,* who encouraged me to be brave and write a book. So, I wrote a book–although different from the book we talked about, which was my biography, a story of self-assurance and great courage against all odds. Maybe someone else will write that one for me. Dave Birch, for instilling confidence in me that I did not actually need a co-author to write a book. It is tough to undertake this vast volume of work, but I am glad that I listened to his advice. I am better at my work because I have done the entire research and wrote the book on my own. Margaret Heffernan for generously sharing her experience as a best-seller author and helping me get on the right track when I started writing this book. Kay Firth-Butterfield, who has always inspired me with her integrity and commitment to ethics in AI. Anne-Marie Durbin, Esther Cavett, and Merrill April for enabling me to reach the next level of my career. I owe you so much. John Matthews for inspiring the original title of this book *Edge of Progress* over a formal dinner in London in 2018. It will be the title of my next (shorter) book. Your achievements in your career continue to inspire me! David Shrier, for his steady advice and guidance to deliver this book. Erik Brynjolfsson, for his thoughtful comments and for his recent research that validated many of the theories presented in this book.

I'm grateful to industry leaders and business executives who I worked with, interviewed for this book, or shared the same stage with at leading fintech conferences across the globe, who shared with me their experience and knowledge showcased in this book. I am also grateful to Yamada-San, for encouraging me to put this book forward to the Japanese readers. I am sincerely grateful that some of these great people have become my friends, too. I am thankful to those organisations which gave me access to their resources and granted me permission to use them in this book. In particular, I am grateful to the National Association of Corporate Directors (NACD) that provided me with unrestricted access and use of some of their recent reports.

Moreover, I am deeply grateful to my best friend, for his saint-like patience and calming guidance during the time of finalising this book. Thank you, James! My son, who has always been by my side—my best teacher, inspiration, and source of wisdom—we are a team. I am very proud of your achievements! My parents, for giving me a strong moral compass of integrity, work ethic, and for believing that I am able to achieve yet another herculean-size task—writing this book. My grandparents, who taught me humility and grounded me in the simplicity of honest values in life.

My professors played a formative role, which has influenced my entire life. My mathematics tutor, a computer scientist, helped me become an A star student with a lifelong love for mathematics. Although I became an economist working in financial services, I've come full circle to use the mathematics he taught me in my current work. My Oxford professors shaped my thinking and helped me to think critically, in particular, Tim Jenkins, Steve New, Colin Mayer, and Maxine Hewitt—who was my grounding force while in Oxford.

And finally, I am grateful to the intellectual community of scholars at Oxford and beyond. Their work forms the basis of many frameworks in this book used to predict the best decision-making processes in the uncharted waters ahead of us. Margaret Boden, Daniel Dennett, Murray Shanahan, and Joanna Bryson are brilliant scholars whose work inspired me to anchor the beginning of this book in philosophical and theological roots.
This book hinges on many discussions I had with the intellectual community in Oxford and on the vast resources the University of Oxford libraries provided during my research, in particular Radcliffe Science Library, Sackler Library, and the computer science library.

I am grateful to my old Oxford college with roots going back to 1282, and which hosted me warmly over the course of more than two months during the early days of my research for my book. I have donated my collection of AI books that I purchased as part of my research to my old college for students to use.

The profits from this book will be donated to two charitable causes:

1. As a bursary to support students in financial distress. Roy Stuart was the Dean of Hertford College during my time in Oxford. Sadly, Roy passed away. His kindness, fairness, and integrity have had a great impact on me. This bursary is in his memory.
2. A contribution to support global reforestations. Each copy of this book funds the planting and caring of one tree to the point at which it can survive on its own. This donation supports the work done by World Land Trust, a charity supported by Sir David Attenborough, Britain's best known and most loved Natural History Filmmaker. Sir David has supported the work of the WLT since its foundation in 1989.

ABOUT THE AUTHOR

Clara Durodié is a technology strategist specialising in ethical artificial intelligence (AI) for business growth and profitability in financial services.

She is internationally recognised for her expertise in corporate governance of AI, strategy, business transformation with AI, and strategic use of data. Clara is an adviser to corporate boards, investment funds, and governments. She is also a mentor, investor, and non-executive director of AI fintech companies. In addition, Clara has advised the World Economic Forum's 4th Industrial Revolution Centre in San Francisco (US), the All Parliamentary Party Group for Artificial Intelligence (UK), and the Ministry of Internal Affairs and Communications' special commission on artificial intelligence (Japan). In June 2018, Clara was also appointed as a member of the European Union Artificial Intelligence Alliance, an EU governmental body.

Clara is a frequent keynote speaker at leading fintech conferences around the globe, including SIBOS, Money 2020, Innovate Finance Global Summit, and AI World. She has been invited as a guest lecturer by leading universities like the University of Oxford's Executive MBA and Massachusetts Institute of Technology. Clara also teaches at the Oxford Artificial Intelligence course, organised by the University of Oxford. In 2015, Clara started scoping the research for her PhD, which sits at the intersection of neuroscience, artificial intelligence, and wealth management. Her focus is on how episodic memory informs how people save and invest and has direct application in AI design in retail banking and wealth management.

In her corporate life before 2014, Clara served in leadership roles in European asset and wealth management in the UK, Switzerland, and Luxembourg. Clara is a member of the Chartered Institute for Securities and Investment (UK), has a Certificate in Investment Management (UK), and holds a Master's degree from the University of Oxford. She is based in London.

Clara can be contacted at the following:

Twitter: @clara_durodie | LinkedIn: Clara Durodie

Send message on *www.cognitivefinance.ai* | *www.clara-durodie.com*

NOTES

Introduction

1. Copyright Clara Durodié. All rights reserved. No permission to use this sentence as it is, without attribution.
2. A play on words using radical transparency, Ray Dalio's core principle to investment management and running Bridgewater Associates, one of the world's largest hedge funds

CHAPTER 1: What Is Intelligence?

3. Faggella, Dan, Two Questions that Humanity Should Be Concerned With, March 11, 2019
4. O'Leary, D., & Kingston, J. (1995). Artificial intelligence in business: Development, integration and organizational issues (AIAI-TR; 166). Edinburgh: Artificial Intelligence Applications Institute, University of Edinburgh.
5. Noumenal from noumenon, which in metaphysics 1. intellectual conception of a thing as it is in itself, not as it is known through perception; 2. The of-itself-unknown and unknowable rational object, or thing-in-itself, which is distinguished from the phenomenon through which it is apprehended by the physical senses, and by which it is interpreted and understood; – so used in the philosophy of Kant and his followers. (source Webster Online dictionary)
6. Howard Gartner identified eight types of intelligence: Linguistic intelligence ("word smart"); Logical-mathematical intelligence ("number/reasoning smart"); Spatial intelligence ("picture smart"); Bodily-Kinesthetic intelligence ("body smart"); Musical intelligence ("music smart"); Interpersonal intelligence ("people smart"); Intrapersonal intelligence ("self smart"); Naturalist intelligence ("nature smart")
7. Constructal law was discovered by Adrian Bejan, a Romanian born MIT physics graduate and US professor. Professor Bejan was awarded the prestigious Benjamin Franklin Medal in 2017 for discovering constructal law. This medal was awarded to leading scientists like Nikola Tesla, Albert Einstein, and Bill Gates. His book Physics of Life is a valuable resource to dive deeper in this theory.

8. Tegmark, Max Life 3.0: Being Human in the Age of Artificial Intelligence. London: Allen Lane, 2017. p. 49

9. Sternberg, R. J. & Salter, W. "Conceptions of intelligence." In Sternberg, R. J. (Ed.), Handbook of Human Intelligence. Cambridge: Cambridge University Press, 1982.

10 Bryson, J., (2018). No one should trust AI – AND – Presenting robots as people stops us thinking clearly about AI

11. Luck, Michael and d'Inverno Mark, "A Formal Framework for Agency and Autonomy", Proceedings from the First International Conference on Multiagent Systems, 1995

12. The Mind Lab, Are You Hallucinating Right Now, an interview with Dr. Anil Seth, September 10, 2018

13 Penrose, R. Shadows of the Mind : A Search for the Missing Science of Consciousness (Corrected ed.). Oxford: Oxford University Press, 1995. p.8

14. Linde, A. D., and Marc. Damashek. Particle Physics and Inflationary Cosmology. Chur; Reading: Harwood Academic, 1990. Print. Contemporary Concepts in Physics, v. 5.

15 Tegmark, M. Life 3.0: "Being Human in the Age of Artificial Intelligence". First ed. New York, 2017. Print.

CHAPTER 2: The Taxonomy of Artificial Intelligence

16. Russell, S., & Norvig, P. (2016). Artificial Intelligence: A Modern Approach (3rd edition.; Global ed., Prentice Hall series in artificial intelligence). Boston; London.

17. idem

18. J. Chalmers, David. (2010). The Singularity: A Philosophical Analysis, Journal of Consciousness Studies. 17. 7-65.

19. Mohamed, Alghanami. (2017). The Role of Artificial Super Intelligence (ASI) in the Energy Sector: Potential, Uncertainties and Barriers.

20. Tegmark, M. (2017). Life 3.0: Being human in the age of artificial intelligence, Penguin series, London.

21. Lauterbach, A., and Bonime-Blanc, A., (2018) The Artificial Intelligence Imperative, Praeger, New York

22. Lovelock, Gartner report 2018

23. Vetter M,. (2018) AI And Machine Learning Will Transform Wealth Management, Forbes

24. David Kelnar, (2019) MMC Ventures The State of AI report, page 31

25. Cormen, T. (2001). Introduction to Algorithms (2nd ed.). Cambridge, Mass.; London: MIT Press. p.11
26. Idem
27. Kaplan, J. (2016). Artificial intelligence: What Everyone Needs to Know. Oxford University Press. p. 27 - 28
28. Elite Data Science, Overfitting in Machine Learning: What It Is and How to Prevent It
29. Silver, D. DeepMind.com, Deep Reinforcement Learning, June 17, 2016.
30. Hassabis, D. and Silver, D. (2017), deepmind.com. AlphaGoZero: Learning from scratch, October 18, 2017
31. Perez, C.E. (2018). Intuition Machines versus Algebraic Minds, Jan 6, 2018
32. Hassabis, D. and Silver, D. (2017). Deepmind.com. AlphaGoZero:Learning from scratch, 18th October 2017
33. Poole, D., & Mackworth, A. (2017). Artificial Intelligence: Foundations of Computational Agents (Second ed.). Cambridge, United Kingdom. p. 552 & p.557
34. Poole, D., & Mackworth, A. (2017). Artificial Intelligence: Foundations of Computational Agents (Second ed.). Cambridge, United Kingdom. p. 552
35. Filos, A. (2018). Reinforcement Learning for Portfolio Management, Department of Electrical Engineering, Imperial College, London
36. Jiang, Z., Xu, D., & Liang, J. (2017). A Deep Reinforcement Learning Framework for the Financial Portfolio Management Problem; Source: Cornell University
37. Pendharkar, Parag C., & Cusatis, Patrick. (2018). Trading financial indices with reinforcement learning agents. Expert Systems With Applications, 103, p. 1-13.
38. Kaplan, J. (2016). Artificial Intelligence: What Everyone Needs to Know. Oxford University Press. p. 27 - 28
39. Clark, J. (2018). Artificial Intelligence: Teaching Machines to Think Like People, O'Reilly Magazine
40. Idem
41. Hinton, G., Sabour, A. and Frost, N. (2018). Research paper, Matrix Capsule with EM Routing.
42. Sara Sabour, Nicholas Frosst, Geoffrey E Hinton, Dynamic Routing Between Capsules, paper 2017
43. Pechyonkin, A. (2017) Understanding Hinton's Capsule Networks, Part 1: Intuition
44. Clark, J., (2017). Adapting ideas from neuroscience for AI, O'Reilly Magazine

45. Gouws, S. (2018). Google Brain Team , Moving beyond translation with the Universal Transformer

46. Creswell, A. et al, (2017). research paper Generative Adversarial Networks: An Overview. IEEE Signal Processing Magazine, vol. 35, issue 1, pp. 53-65

47. Goodfellow, I., Pouget-Abadie, J., Mirza, M., Xu, B., Warde-Farley, D., Ozair, S.,Bengio, Y. (2014). Generative Adversarial Networks.

48. Cholaquidis et al. (2018). Semi-supervised learning: when and why it works? joint research paper Universidad de la Republica, Uruguay and Universidad de Buenos Aires

49. Semwal, T. et al. (2018) research paper A Practitioners' Guide to Transfer Learning for Text Classification Using Convolutional Neural Networks

50. Ruder, S., (2019). Neural Transfer Learning for Natural Language Processing (PhD thesis)

51. Mesnil, G. et al. (2011) research paper Unsupervised and Transfer Learning Challenge: a Deep Learning Approach

52. Torrey, L and Shavlik, J. (2009). Transfer Learning. In Handbook of Research on Machine Learning Applications and Trends: Algorithms, Methods, and Techniques, IGI Global

53. Sousa, R. et al. (2014). research paper Transfer Learning: Current Status, Trends and Challenges

54. Mcmahan, B. and Ramange, D. (2017). Google AI Blog, "Federated learning" April 6, 2017

55. Intel AI (2019), Federated Learning for Medical Research

56. de Vinzelles, G., (2018). "Federated Learning: a step closer towards confidential Artificial Intelligence"

57. Smith, V. et al. (2018). Research paper Federated Multi-tasking Learning

58. Anguita, D. et al (2013) "A public domain dataset for human activity recognition using smartphones." In European Symposium on Artificial Neural Networks, Computational Intelligence and Machine Learning

59. Pantelopoulos, A. and Bourbakis N.G. (2010) "A survey on wearable sensor-based systems for health monitoring and prognosis." IEEE Transactions on Systems, Man, and Cybernetics, 40(1): 1–12

60. Rashidi, P. and Cook, D. J. (2009). "Keeping the resident in the loop: Adapting the smart home to the user." IEEE Transactions on systems, man, and cybernetics, 39(5): 949–959

61. Simard et al. (2017). Research Paper Machine teaching: a new paradigm for building machine learning systems

62. IntelAI, nGraph: Focus on Data Science, not on Machine Code

63. Lauterbach, A., Bonime-Blanc, A., (2018) "The AI Imperative", Praeger

64. Sastry K., Goldberg D., Kendall G. (2005) "Genetic Algorithms." In: Burke E.K., Kendall G. (eds) Search Methodologies. Springer, Boston, MA

65. Russo, C. "New Twist on AI Evolutionary Algorithms in Neuroscience." Psyschology Today. May 24, 2019 accessed 12 August 2019

66. Kroha P., Friedrich M. (2014) "Comparison of Genetic Algorithms for Trading Strategies." In: Geffert V., Preneel B., Rovan B., Štuller J., Tjoa A.M. (eds) SOFSEM 2014: Theory and Practice of Computer Science. SOFSEM 2014. Lecture Notes in Computer Science, vol 8327. Springer, Cham https://link.springer.com/chapter/10.1007/978-3-319-04298-5_34

67. Google AI, Research Teams, Language.

68. Le, J., (2019). The Five Trends Dominating Computer Vision

69. Finextra, (2018). Is the Financial Sector Ready for Innovation with Computer Vision?

70. Nilsson, N. (2010). The Quest for Artificial Intelligence: A History of Ideas and Achievements. Cambridge: Cambridge University Press.

71. Hurwitz, et al. 2015

72. Lauterbach, A. and Bonime-Blanc. (2018). The AI Imperative. Praeger

73. Ferrucci, D. (2012). Introduction to "This is Watson." IBM Journal Of Research And Development, 56(3-4), 1:1-1:15

74. Ashton, K., (2009) That 'Internet of Things' Thing, In the real world, things matter more than idea

75. Daecher, A. and Galizia, T., Deloitte Insights, Ambient Computing: Putting the Internet of Things to Work, 2015

CHAPTER 3: Prerequisites for Artificial Intelligence

76. Holmes, D. (2017). Big data: A very short introduction (Very short introductions; 539). Oxford. p.3

77. Idem
78. Dell company website. Global Technology Adoption Index 2015"https://blog.dell.com/en-us/global-technology-adoption-index-2015/ (accessed on 19th Aug 2018)
79. Lauterbach, A. and Bonime-Blanc, A. (2018) The AI Imperative, p. 61, Praeger, New York
80. Laney, D. (2001), Meta Group. 3D Data Management: Controlling data Volume, Velocity and Variety
81. IBM Big Data Hub, Infographic The Four V's of Big Data
82. Marr, B. (2017) Data Strategy, p. 87, Kogan International, London
83. Robb, D. (2017). Semi-Structured Data, Datamation
84. The Financial Times, (2018) Asset managers double spending on new data on hunt for edge.
85. Baert, N, April 2018, Pension & Investments "Location data provides new insights into investments
86. Image-Net website accessed on 20 August 2018
87. Halevy, A., Norvig, P., & Pereira, F. (2009). "The Unreasonable Effectiveness of Data." Intelligent Systems, IEEE, 24(2), 8-12.
88. Tawakol, O. (2012) More Data Beats Better Algorithms – Or Does It?
89. AIMultiple webpage, (2019). Synthetic Data: An Introduction and 10 Tools
90. Amazon website, Data Lakes and Analytics on AWS, What is a Data Lake ?
91.Sachdeva, K., (2017). Incorporating machine learning in the data lake for robust business results
92. Ali, M., LionBridge article (2019) "The 50 Best Free Datasets for Machine Learning"
93. This is an aggregated source from (1) https://medium.com/startup-grind/fueling-the-ai-gold-rush-7ae438505bc2; (2) https://www.forbes.com/sites/bernardmarr/2018/02/26/big-data-and-ai-30-amazing-and-free-public-data-sources-for-2018/#7d4a42925f8a and own research
94. Tang, C. (2016). The data industry: The Business and Economics of Information and Big Data. (Ebook central). Hoboken, New Jersey.
95. CBInsights, August 2018, 7 Companies Using Blockchain To Power AI Applications
96. Yao, M. (2017). "You can't build enterprise AI if you suck at data & analytics. " CIO Magazine

97. Lauterbach, A. and Bonime-Blanc, A., (2018). The AI Imperative. p. 66, Praeger, New York
98. Google AI, The Google File System, https://ai.google/research/pubs/pub51
99. Vance, Ashely, (2009). Hadoop, a Free Software Program, Finds Uses Beyond Search, March 16, 2009
100. The Spark Apache Foundation, https://spark.apache.org
101. Idem
102. Amazon Web Services company website.Article: Big Data, What is Presto
103. Dignan, L. (2018) Cloudera, Hortonworks merge in deal valued at $5.2 billion, October 3rd, 2018
104. Goldman Sachs Asset Management Perspectives (2016). The role of big data in investments Roundtable https://www.gsam.com/content/gsam/us/en/individual/market-insights/gsam-insights/gsam-perspectives/2016/big-data/gsam-roundtable.html
105. Idem
106. This table draws from (1) Lauterbach and Bonime-Blanc, 2018, The AI Imperative, p. 66 and from (2) Owler companies information database
107. Dong, G., & Liu, H. (2018). Feature Engineering for Machine Learning and Data Analytics (1st ed., Chapman & Hall/CRC Data Mining and Knowledge Discovery Series).
108. NaraLogics company website https://naralogics.com accessed on 28 August 2018
109. IDC research, 25 September 2017, IDC Spending Guide Forecasts Worldwide Spending on Cognitive and Artificial Intelligence Systems to Reach $57.6 Billion in 2021
110. Right Scale survey, 2018, State of the Cloud Report™, 2018 Data to Navigate Your Multi-Cloud Strategy
111. Harvey, C., (2018) Artificial Intelligence as a Service: AI Meets the Cloud, posted March 15, 2018
112. MIT News Offices, Explained: The Shannon Limit
113. Koch et al. Federal University of Santa Catarina, Florianopolis, Brazil and Western Santa Catarina State University, Videira, Brazil. "Distributed artificial intelligence for network management - new approaches, Springer
114. Greenmeier, L., (2013) When Will the Internet Reach Its Limit (and How Do We Stop That from Happening)? article in The Scientific American Magazine
115. The Computer History Museum, Time line of Computer History, (2012)

116. amazon.co.uk search for '1 GB memory card' listed price £13 for SanDisk 1 GB digital SD memory card(SDSDB-1024-AW110) accessed on 27 August 2018
117. Bank of America Merrill Lynch (2016) Global Semiconductors: Deep Learning and the processor chips fueling the AI revolution – a primer
118. Moore, S., (1995) The Coming Age of Abundance in Ronald Bailey, ed., The True State of the Planet, Free Press, 1995, p.113
119. Intel company website, From Sand to Circuits, The surprising process behind Intel® technology
120. Anthes, G., How To Making Microchips, (2002) article in ComputerWorld
121. Wikipedia, Integrated Circuit
122. Lauterbach A. and Bonime-Blanc, A., The AI Imperative, 2018, Praeger, New York
123. Schlegel, D., (2015) University of Heidelberg , Deep Machine Learning on GPUs
124. Algorithmia company website (2018) Hardware for Machine Learning
125. Schlegel, D., (2015) University of Heidelberg , Deep Machine Learning on GPUs
126. Google Cloud, Cloud TPU. Train and run machine learning models faster than ever before
127. Codon, S., (2017) TPU is 15x to 30x faster than GPUs and CPUs, Google says, article in ZDNet, April 5, 2017
128. Miller, R., (2017) Google's second generation TPU chips takes machine learning processing to a new level, article in TechCrunch May 17, 2017
129. Mannes, J., (2017) Google is giving a cluster of 1,000 Cloud TPUs to researchers for free article in TechCrunch May 17, 2017
130. Cutress, I., Cambricon, Makers of Huawei's Kirin NPU IP, Build A Big AI Chip and PCIe Card article in AnandTech
131. Xilinx company website, Field Programmable Gate Array
132. W. Miller, "Real world applications for field programmable gate array devices-an overview," Proceedings of WESCON '94, Anaheim, CA, USA, 1994, pp. 548-551.
133. Freund, K., A Machine Learning Landscape: Where AMD, Intel, NVIDIA, Qualcomm And Xilinx AI Engines Live article in Forbes Marc 3, 2017
134. Simonite, T., Apple's 'Neural Engine' Infuses the iPhone With AI, article in Wired September 13, 2017
135. Intel Newsroom, (2018) Intel and Microsoft Enable AI Inference at the Edge with Intel Movidius Vision Processing Units on Windows ML

136. IntelAI website, Nervana NNP
137. Harris, N., Lightmatter+GV post on medium.com Feb 25 2019 accessed 26 February 2019
138. Lauterbach and Bonime-Blanc, 2018, The AI Imperative p.64, Preaeger, New York
139. Designer's Guide: Selecting AI chips for embedded designs posted on Electronic Products
140. Paxata company website, Best Practices for Creating and Operationalizing Data Lakes

CHAPTER 4: The Current State of AI—Evolution, Growth, and Investing

141. I agree with this detailed cycle as suggested by David Shrier
142. DRAPA TV (2017), A DARPA Perspective on Artificial Intelligence (video)
143. CrunchBase (2019) Q4 2018 Closes Out A Record Year For The Global VC Market
144. CB Insights Research (2019) Venture Capital Funding Report 2018
145.Vilar, H., (2019) MUFG launches new $185m fintech fund; article in Bankingtech,
146. Lee, T. and Gurman, D. (2018) 10 Best Practices For Due Diligence In AI Transactions
147. Zia, C. (2018) Artificial intelligence: winter is coming. Article published in the Financial Times October 17 2018
148. The FT, Ram, A., March 5 2019, Europe's AI start-ups often do not use AI, study finds
149. Bank Of England, (2019) Embracing the promise of fintech, Quartely Bulletin 2019 Q1
150. idem
151. Bloomberg, "Meet the German Fintech That's Now Worth More Than Deutsche Bank," by Geoffrey Smith, 14 August 2018
152. Oaknorth company information provided June 2018, reproduced with permission
153. The Guardian, Patrick Collison, August 12, 2019, "Hey Google, lend me tenner? NatWest trials voice banking."
154. IPA, APS, IA, AIA, DI stand for respectively Intelligent Process Automation, Augmented Process Stream, Intelligent Automation, Augmented Intelligent Automation, Digital Intelligence
155. Shearman and Sterling, " Fintech Regulatory SandBoxes" updated May 2019
156. Bloomberg Companies Database, SenseTime Group Ltd.
157. ToughtMachine company website https://www.thoughtmachine.net/vault/

158. Lloyds Banking Group's press release (2018) "Lloyds Banking Group Enters Into Strategic Partnership with Thoughts Machine" 7 November 2019
159. as of 11 April 2019
160. Allen, Gregory. "Understanding China's AI Strategy". Center for a New American Security. Center for a New American Security. Retrieved 11 April 2019.
161. The Economist article "Artificial intelligence is awakening the chip industry's animal spirits."Jun 7th 2018
162. Semiconductors are often referred to as chips or silicon.
163. Algorithmia company website. "Hardware for Machine Learning" accessed 12 March 2019
164. Burridge, N."Artificial intelligence gets a seat in the boardroom" article in Nikkei Asian Review May 10, 2017
165. McKinsey, "A machine-learning approach to venture capital" June 2017
166. Federoff, S., "Manuela Veloso departs CMU for J.P. Morgan" Pittsburgh Business TimesMay 7, 2018
167. idem
168. JP Morgan company website, "J.P. Morgan AI Research"
169. The Financial Times, John Thornhill August 14, 2015, Lunch with the FT: Mariana Mazzucato, accessed 16 November 2018
170. The New York Times "Beijing Wants A.I. to Be Made in China by 2030"

CHAPTER 5: Corporate Governance and AI Adoption

171. This is attributed to Noel Sharkey PhD, Emeritus Professor of AI and Robotics at University of Sheffield (UK), co-director Responsible Robotics, Chair of ICRAC. His current core research interest is in ethical applications of robotics and AI in areas such as the military, child care, elder care, policing, surveillance, medicine/surgery, education and criminal/ terrorist activity.
172.World Economic Forum, The Value of Data, article published 22 September 2019
173. Quote attributed to Daniel Radclyffe, a digital ethicist and AI programme director at Fidelity International.
174. idem
175. The Future of Humanity Institute, University of Oxford, Centre for the Governance of Artificial Intelligence

176. https://twitter.com/coe/status/1093405085978243077?s=21 accessed on 10 February 2019
177. The Telegraph, Jessica Carpani, 19 June 2019, "Oxford University given £150m by US billionaire to investigate AI in biggest ever donation."
178. https://www.umu.se/en/news/umea-university-to-lead-prestigious-ai-program_7983837/
179. Sara Baase, "The gift of fire, social, legal, and ethical issues for computing technology, 4th edition. Pearson Education, 2013, p.44
180. Deborah G. Johnson, Computer ethics, Prentice Hall, 2nd ed., 1994
181. Sara Baase, The gift of fire, social, legal, and ethical issues for computing technology, 4th edition. Pearson Education, 2013
182. ACM/IEEE-CS Software Engineering Code accessed on 16 November 2018
183. R. L. Glass, "The state of the practice of software engineering," in IEEE Software, vol. 20, no. 6, pp. 20-21, Nov.-Dec. 2003. doi: 10.1109/MS.2003.1241361
184. The Future Computed: Artificial Intelligence and its role in society, Microsoft, https://news.microsoft.com/uploads/2018/02/The-Future-Computed_2.8.18.pdf accessed 17 November 2018
185. https://ethicalos.org
186. Fred Wilson, daily newsletter 20th March 2019
187. Alan Winfield Blog, "An Updated Roundup of Ethical Principles of Robotics and AI" April 18, 2019
188. European Commision, "High-Level Expert Group on Artificial Intelligence"
189. Beresford, Tom. "Algorithmic transparency is not the solution you're looking for – algorithmic accountability is." Gamification of Work, 2 November 2016.
190. Federal register, National Archives, request for information on Artificial Intelligence, one 26, 2016; accessed on 17 November 2018
191. Muller, V., Bostrom, N. 'Future progress in artificial intelligence: A Survey of Expert Opinion, in Vincent C. Müller (ed.), Fundamental Issues of Artificial Intelligence (Synthese Library; Berlin: Springer)
192. https://arxiv.org/abs/1606.06565 accessed on 18 November 2018
193. Cognitive Scale company website. Press release "CognitiveScale Announces Cortex Certifai to Tackle Trust as a Barrier to Enterprise AI Adoption"

194. Carnegie Mellon University, School of Computer Science, B.S. IN ARTIFICIAL INTELLIGENCE Curriculum
195. Joy Buolamwini, J., "How does facial recognition software see skin color?", MIT Media Lab, 2018
196. Statistica, Overfitting accessed October 15, 2019
197. Suresh, H. and Guttag, J. "A Framework for Understanding Unintended Consequences of Machine Learning"
198. Bryson, J., Joanna Bryson Blogspot, "Three Very Different Sources of Bias"
199. idem
200. Cathy O'Neil - The era of blind faith in big data must end - TED talk 2017
201. Cathy O'Neil - The era of blind faith in big data must end - TED talk 2017
202. Campbell-Dolaghan, K., "The Art Of Manipulating Algorithms" article in Fast Company 03.01.17
203. Y Combinator website "At the Intersection of AI, Governments, and Google – Tim Hwang"
204. Hao, K. "Congress wants to protect you from biased algorithms, deepfakes, and other bad AI" article in MIT Technology Magazine
205. Professor Frank Pasquale's personal website
206. Zammuto, M., " In the AI revolution, bias is the new breach. How CIO's must manage risk" article in CIO Magazine
207. Professor Joanna Bryson, 2019
208. Makkula Center for Applied Ethics at Santa Clara University "What is Privacy?"
209. The Economist, The world in 2019 issue - accessed on The Economist app on 24 November 2018.
210. Meeco company website www.meeco.me
211. BBC News technology, Leo Kelion, "MiSafes' child-tracking smartwatches are 'easy to hack" https://www.bbc.co.uk/news/technology-46195189
212. Francis X. Shen, University of Minnesota"Sex robots are here, but laws aren't keeping up with the ethical and privacy issues they raise" https://themoderatevoice.com/sex-robots-are-here-but-laws-arent-keeping-up-with-the-ethical-and-privacy-issues-they-raise/
213. Privacy International "Tell companies to stop exploiting your data!" https://privacyinternational.org/campaigns/tell-companies-stop-exploiting-your-data
214. The EU General Data Protection Regulation (GDPR) Article 35(1)

215. De Montjoye, Y., Radaelli, L., Singh, V., & Pentland, A. (2015). "Identity and privacy. Unique in the shopping mall: On the reidentifiability of credit card metadata." Science (New York, N.Y.), 347(6221), 536-9.
216. Information Commissioner's Office. Conducting privacy impact assessments code of practice. ICO, February 2014
217. Big data, artificial intelligence, machine learning and data protection 20170904 Version: 2.2 https://ico.org.uk/media/for-organisations/documents/2013559/big-data-ai-ml-and-data-protection.pdf, Annex 1 – Privacy impact assessments for big data analytics, page 99-113, accessed 12 November 2018.
218. Law 360 "Cybersecurity & Privacy Predictions For 2019"January 1, 2019 https://www.law360.com/articles/1112115/cybersecurity-privacy-predictions-for-2019 accessed 18 Jan, 2019
219. Apple vows to resist FBI demand to crack iPhone linked to San Bernardino, accessed 18 Jan 2019
220. IEEE, Ramaswamy Palaniappan ; Danilo P. Mandic "Biometrics from Brain Electrical Activity: A Machine Learning Approach" https://ieeexplore.ieee.org/abstract/document/4107575
221.Biometric, Chris Burt, "Fujitsu palm vein tech to be tested for cardless retail payments in Japan" https://www.biometricupdate.com/201807/fujitsu-palm-vein-authentication-technology-to-be-tested-for-cardless-retail-payment-in-japan
222. Stephen Mayhew, "Hitachi, KDDI testing blockchain system for retail payments using finger veins" https://www.biometricupdate.com/201807/hitachi-kddi-testing-blockchain-system-for-retail-payments-using-finger-veins
223. Goode Intelligence Report "Biometrics for Payments; Market and Technology Analysis, Adoption Strategies and Forecasts 2018-2023 – Second Edition″
224. Law 360, "Sekura Case Expands Scope Of Illinois Biometric Privacy Law" November 14, 2018 https://www.law360.com/illinois/articles/1100521/sekura-case-expands-scope-of-illinois-biometric-privacy-law
225.Cas Proffitt "The Future Of Biometrics With Artificial Intelligence" 9 August 2017 https://www.disruptordaily.com/future-biometrics-artificial-intelligence/
226.CB Insights companies database: BioCatch https://www.cbinsights.com/company/biocatch

227. Forbes, Jonathan Vanian"Artificial Intelligence Is Giving Rise to Fake Fingerprints. Here's Why You Should Be Worried"November 28, 2018 http://fortune.com/2018/11/28/artificial-intelligence-fingerprints-security/
228. Tero Karras "A Style-Based Generator Architecture for Generative Adversarial Networks" 29 March 2019 https://arxiv.org/pdf/1812.04948.pdf
229. CB Insights, Future of Information Warfare accessed 16 February 2019 https://www.cbinsights.com/research/new-report-information-warfare/
230. idem
231. Wired, Lilly Hay Newman,"Machine Learning Can Create Fake 'Master Key' Fingerprints" 11.17.2018 https://www.wired.com/story/deepmasterprints-fake-fingerprints-machine-learning/
232. Truepic company website https://truepic.com/technology/
233. 2018 "Antitrust and Competition" Conference, https://research.chicagobooth.edu/stigler/events/single-events/antitrust-competition-conference-digital-platforms-concentration?mc_cid=ab6c328a3e&mc_eid=62e547fefc
234. Carr Ferrell LLP, Gary Rebackhttp://www.carrferrell.com/attorneys/gary-reback
235. "Digital Platforms and Concentration" Conference https://www.youtube.com/watch?v=qzjfjYQTa2s&feature=youtu.be at minute 16:12; accessed on 20 November 2018
236. The New York Times, "Google Fined Record $2.7 Billion in E.U. Antitrust Ruling" https://www.nytimes.com/2017/06/27/technology/eu-google-fine.html accessed 18 November 2018
237. "Digital Platforms and Concentration" Conference https://www.youtube.com/watch?v=qzjfjYQTa2s&feature=youtu.be accessed on 20 November 2018
238. The Future of Life Institute "Artificial Intelligence and Income Inequality" March 16, 2017. https://futureoflife.org/2017/03/16/shared-prosperity-principle/ accessed 24 November 2018.
239. The Future of Life Institute, The ASILOMAR AI PRINCIPLES https://futureoflife.org/ai-principles/
240. Forbes, Maribel Lopez, "MIT And IBM Lab Partnership Launches 48 Projects To Tackle AI Challenges" September 26, 2018. https://www.forbes.com/sites/maribellopez/2018/09/26/mit-and-ibm-lab-partnership-launches-48-projects-to-tackle-ai-challenges/#404b47ab97d2

241. Rob Bensinger, "Stuart Russell: AI value alignment problem must be an ″intrinsic part″ of the field′s mainstream agenda" https://www.lesswrong.com/posts/S95qCHBXtASmYyGSs/stuart-russell-ai-value-alignment-problem-must-be-an
242. Monetary Authority of Singapore, "Principles to Promote Fairness, Ethics, Accountability and Transparency (FEAT) in the Use of Artificial Intelligence and Data Analytics in Singapore's Financial Sector". http://www.mas.gov.sg/~/media/MAS/News%20and%20Publications/Monographs%20and%20Information%20Papers/FEAT%20Principles%20Final.pdf
243.Deloitte, "State of AI in the Enterprise, 2nd Edition" 22 October, 2018. https://www2.deloitte.com/insights/us/en/focus/cognitive-technologies/state-of-ai-and-intelligent-automation-in-business-survey.html
244. Explainable AI Accelerates Digital Transformation in the Financial Services Industry, Enhancing Risk Management and Other Operations. https://journal.jp.fujitsu.com/en/2018/10/11/01/ accessed 20 February 2019
245. Makridakis, S (2017) "The forthcoming artificial intelligence revolution: its impact on society and firms." Futures, 90, 46-60
246 Clara Durodie was invited in 2018 on behalf of Japan′s Ministry of Internal Affairs and communication to provide her input and comments on these principles.
247. CNN Business, Heather Kelly, "AI is hurting people of color and the poor. Experts want to fix that" July 23, 2018. https://money.cnn.com/2018/07/23/technology/ai-bias-future/index.html
248. Wooldridge, M., Artificial Intelligence, Ladybird Expert series, 2018

CHAPTER 6: Boardroom Oversight and AI Adoption

249.BetterBoards company website www.better-boards.com
250. BCG "The Art of Risk Management" APRIL 30, 2017. https://www.bcg.com/publications/2017/finance-function-excellence-corporate-development-art-risk-management.aspx
251. McKinsey, "Building a forward-looking board" https://www.mckinsey.com/business-functions/strategy-and-corporate-finance/our-insights/building-a-forward-looking-board
252. Summer 2018 FT-ISCA Ballroom Bellwether Survey https://www.icsa.org.uk/knowledge/research/ft-icsa-boardroom-bellwether-survey-summer-2018

253. MITSloan Management Review, Thomas H. Davenport and Vivek Katyal "Every Leader's Guide to the Ethics of AI" December 06, 2018. https://sloanreview.mit.edu/article/every-leaders-guide-to-the-ethics-of-ai/
254. NACD and Protiviti, Is Board Oversight Addressing the Right Risks? Strategies for Addressing the New Risk Landscape (2018), page 16.
255. NACD and Protiviti, Is Board Oversight Addressing the Right Risks? Strategies for Addressing the New Risk Landscape (2018), page 15.
256. The Report of the NACD Blue Ribbon Commission on Adaptive Governance: Board Oversight of Disruptive Risks (Arlington, VA: NACD, 2018), pages 19–20.
257. Data from NACD member poll on board oversight of atypical risk, conducted via email, March–April 2018.
258. The Report of the NACD Blue Ribbon Commission on Adaptive Governance: Board Oversight of Disruptive Risks (Arlington, VA: NACD, 2018), pages 15–17
259. The Report of the NACD Blue Ribbon Commission on Adaptive Governance: Board Oversight of Disruptive Risks (Arlington, VA: NACD, 2018), page 17
260. Anonymous. (2017). FSB report examines the risks and benefits of AI and machine learning in financial services. Hedge Funds and Private Equity, 11(9), 6-7; http://www.fsb.org/2017/11/artificial-intelligence-and-machine-learning-in-financial-service/
261. McKinsey, "Building a forward-looking board" https://www.mckinsey.com/business-functions/strategy-and-corporate-finance/our-insights/building-a-forward-looking-board
262. Diligent company website "What is the Diligent Governance Cloud?" https://diligent.com/en-gb/blog/what-is-the-governance-cloud/ accessed 19 September 2019
263. Board Intelligence company website, "Information Overload Is Crippling Board Performance" 7 March 2019. https://www.boardintelligence.com/blog/information-overload-is-crippling-board-performance
264. Möslein, Florian. "Robots in the Boardroom: Artificial Intelligence and Corporate Law (September 15, 2017)." in: Woodrow Barfield and Ugo Pagallo (eds), Research Handbook on the Law of Artificial Intelligence, Edward Elgar, (2017/18, Forthcoming) . Available at SSRN: https://ssrn.com/abstract=3037403 or http://dx.doi.org/10.2139/ssrn.3037403
265. Dietterich, T.G., Machine learning for real-time decision making, 2001

266. Business Insider, "How Salesforce CEO Marc Benioff uses artificial intelligence to end internal politics at meetings" May 19, 2017. https://www.businessinsider.com/benioff-uses-ai-to-end-politics-at-staff-meetings-2017-5
267. Popoyan L., Napoletano M., and Roventini, A., 14 December 2015, "Taming macroeconomic instability: monetary and macro prudential policy interactions in an agent-based model" http://www.isigrowth.eu/2015/12/14/taming-macroeconomic-instability-monetary-and-macro-prudential-policy-interactions-in-an-agent-based-model/
268. Irish Management Institute, "Bringing Data Modelling into the Boardroom" 5 May 2017. https://www.imi.ie/bringing-data-modelling-boardroom
269. MIT Sloan Management Review, Peter Weill, Thomas Apel, Stephanie L. Woerner, and Jennifer S. Banner "It Pays to Have a Digitally Savvy Board" Spring 2019 Issue, https://sloanreview.mit.edu/article/it-pays-to-have-a-digitally-savvy-board/amp/
270. The shell company concept is attributed to Professor Joanna Bryson.
271. Harvard Law School Forum on Corporate Governance and Financial Regulation "Selecting Directors Using Machine Learning" Michael S. Weisbach (The Ohio State University), on Monday, April 9, 2018. https://corpgov.law.harvard.edu/2018/04/09/selecting-directors-using-machine-learning/
272. MIT Sloan Management Review, "How unprepared are leaders for digital transformation: the blind spots in adopting digital innovation" April 19, 2019
273. Stafford, Brian and Schindlinger, Dottie (2019). Governance in the Digital Age: A Guide for the Modern Corporate Board Director, (pp. 125-126). New Jersey: John Wiley & Sons, Publishers, ©2019, Diligent Corporation. The framework was reproduced with authors' and Wiley's express permission. No part of it may be reproduced, sorted in a retrieval system, or transmitted in any form or by any means, electronic, mechanical, photocopying, recording, scanning or otherwise, except as permitted under Section 107 and 108 of the 1976 United States Copyright Act. For full limitations see Wiley's website.

CHAPTER 7: Strategic Adoption of AI in Business

274. Microsoft Research, "Maximising the AI Opportunity, How to harness the potential of AI effectively and ethically". https://info.microsoft.com
275. AXA Investment Managers, The Thomas Report. November 2018.

276. Disclosure: Clara Durodié is a member of the teaching team at Oxford Artificial Intelligence course. In 2019 Times Higher Education named the University of Oxford the best in the world for computer science courses, ahead of Stanford and MIT.
277. MIT Digital, "Q&A: How AI Can Lead Corporate Strategy" June 24, 2019. https://medium.com/mit-initiative-on-the-digital-economy/q-a-how-ai-can-lead-corporate-strategy-87c7a7561318
278. Kieren McCarthy "Accenture sued over website redesign so bad it Hertz: Car hire biz demands $32m+ for 'defective' cyber-revamp" 23 Apr 2019
279. Brynjolfsson, E., Hitt, L., & Yang, S. (2002). "Intangible assets: Computers and organizational capital." Brookings Papers On Economic Activity, (1), 137-198.
280. Brynjolfsson, E., Hitt, L., 2003, "Computing productivity: firm-level evidence." The Review of Economics and Statistics, November 2003, 85(4): 793–808; © 2003 by the President and Fellows of Harvard College and the Massachusetts Institute of Technology
281. Tesla Autonomy Investor Day, April 22, 2019
282. Harvard Business School Rajiv Lal and Lisa Mazzanti "Goldman Sachs: Anchoring Standards After the Financial Crisis" May 2014 issue https://www.hbs.edu/faculty/Pages/item.aspx?num=47344
283. The Independent Ireland, Donal O'Donovan"Goldman Sachs backs deal to buy GE Capital's Irish loans" August 27 2012. https://www.independent.ie/business/irish/goldman-sachs-backs-deal-to-buy-ge-capitals-irish-loans-26891236.html
284 "Goldman Sachs Bank USA to Acquire the on line deposit platform and assume the deposits of GE Capital" August 13, 2015. https://www.goldmansachs.com/media-relations/press-releases/current/announcement-13-aug-2015.html
285 S2E Transformation company page "Top Frequently asked business architecture questions" https://www.s2etransformation.com/20-frequently-asked-business-architecture-questions/
286. idem
287. "Joachim Wuermeling: Artificial intelligence (AI) in finance - six warnings from a central banker" https://www.bis.org/review/r180307d.pdf
288. GainX company information and website
289. SAS company website "The 5 Essential Components of a Data Strategy" https://www.sas.com/content/dam/SAS/en_us/doc/whitepaper1/5-essential-components-of-data-strategy-108109.pdf

290. Brynjolfsson E, Rock D, Mitchel T, What Can Machines Learn, 2018
291. Harvard Business Review, "The AI Roles Some Companies Forget to Fill", Megan Beck,Thomas H. Davenport, Barry Libert, March 14, 2019.
292. Marr, Bernard. (2017). Data Strategy, Kogan Page. p. 18
293. I came across this in statement during my research for Decoding AI in Financial Service. Unfortunately, I couldn't not find its rightful author. If the author reads it, please get in touch so I can give the right credit.
294. Abraham C., Sims R., Daultrey S., Buff A., Really A., "How digital trust drives culture change," March 18, 2019, MIT Sloan Management Review
295. Idem
296. Brett King, Podcast Breaking Banks, Dave Birch joins to host. Guests include Ruth Wandhofer, Global Head of Regulatory at Citi; Mike Blalock, General Manager of Intel FSI; Poppy Gustafsson, EMEA CEO of Darktrace; Will Beeson, CEO of Civilised Bank and host of ReBank Podcast; Paul Taylor, CEO of Thought Machine; and Kosta Peric from the Bill & Melinda Gates Foundation. https://soundcloud.com/breakingbanks/the-innovate-finance-global-summit, minute 5
297. Harvard Business Review, Thomas H. Davenport, Shivaji Dasgupta "How to Set Up an AI Center of Excellence" January 16, 2019. https://hbr.org/2019/01/how-to-set-up-an-ai-center-of-excellence

CHAPTER 8: How AI Is Reshaping Financial Services

298. https://www.weforum.org/events/world-economic-forum-annual-meeting
299. "Robots in Finance Bring New Risks to Stability, Regulators Warn. Silla Brush," Bloomberg Magazine. 1 Nov 2017
300. Buchanan, Bonnie. (2019, April 02). "Artificial intelligence in finance." Zenodo. http://doi.org/10.5281/zenodo.2612537
301. The Financial Times, "Market pressure blurs the line between US asset and wealth managers," 15th March 2019, Gabriel Altbach
302. The Financial Times, "Jupiter's chief Andrew Formica faces first of many challenges," 5th July 2019, Owen Walker
303. Wealth Management, "Larry Fink on the U.S. Economy and Future of Financial Advice" Michael Thrasher, May 01, 2017. https://www.wealthmanagement.com/industry/larry-fink-us-economy-and-future-financial-advice

304. Financial Conduct Authority Asset Management Market Study, June 2017 https://www.fca.org.uk/publication/market-studies/ms15-2-3.pdf
305. Good Banque, "Qingdao asserts itself as an international asset management center" 07.06.2019. https://goodbanque.com/index.php/2019/07/06/qingdao-asserts-itself-as-an-international-asset-management-center/13966/
306. Dan Mikulskis Investment Partner at LCP "Better Behavior by Design" March 28, 2019. https://www.linkedin.com/pulse/better-behavior-design-dan-mikulskis
307. EConsultancy, Adobe and Pateek Vatash, 2018 report Digital Trends in Financial Services. https://wwwimages2.adobe.com/content/dam/acom/uk/modal-offers/pdfs/Econsultancy-2018-Digital-Trends-FS_EMEA.pdf
308. The Wall Street Journal, Asjylyn Loder "Vanguard Ups the Ante in an ETF Race to Zero" Feb. 26, 2019https://www.wsj.com/articles/vanguard-ups-the-ante-in-an-etf-race-to-zero-11551184467
309. Financial Planning, "Fidelity cuts some fees to zero in price war, stinging BlackRock" August 01, 2018 https://www.financial-planning.com/articles/fidelity-cuts-some-fees-to-zero-in-price-war-stinging-blackrock
310. The Wall Street Journal, Asjylyn Loder "Vanguard Ups the Ante in an ETF Race to Zero" Feb. 26, 2019https://www.wsj.com/articles/vanguard-ups-the-ante-in-an-etf-race-to-zero-11551184467
311. MIT Management Sloan School, Tom Relihan ,"Diving deep into the digital economy" Feb 22, 2019 http://mitsloan.mit.edu/ideas-made-to-matter/diving-deep-digital-economy
312. Capuzzi, B., "The finite of advice is still human," Investment News, 29 August 2019
313. EY, "How do you build value when clients want more than wealth? 2019 Global Wealth Management Research Report"
314. The Wall Street Journal, Yuka Hayashi, Friday May 10th May 2019
315. Forbes, Moira Vetter "AI And Machine Learning Will Transform Wealth Management" https://www-forbes-com.cdn.ampproject.org/c/s/www.forbes.com/sites/moiravetter/2018/07/18/ai-and-machine-learning-will-transform-wealth-management/amp/
316. Backend Benchmark, The Robo Report - second quarter 2018 (https://theroboreport.com)https://st,orage.googleapis.com/gcs-wp.theroboreport.com/gZhdpxRTPEhtB8Wj/2Q%202018%20Robo%20Report.pdf

317. "EY FinTech Adoption Index 2017: The rapid emergence of FinTech" http://www.kurzweilai.net/images/FOLIO-by-Ernst-+-Young-fin-tech-adoption-index.pdf http://www.kurzweilai.net/digest-the-digital-way-forward-for-finance

318. Apple company page "Introducing Apple Card, a new kind of credit card created by Apple" https://www.apple.com/newsroom/2019/03/introducing-apple-card-a-new-kind-of-credit-card-created-by-apple/

319. Oaknorth 2018 company information - published with permission

320. Nav company page. https://www.nav.com accessed 26 February 2019

321. Denise Garth "Insurance at the Intersections of Protection and Prevention" https://coverager.com/insurance-at-the-intersections-of-protection-and-prevention/

322. CBInsights Research, "What's next in Insurance", briefing notes 2019

323. Denise Garth "Insurance at the Intersections of Protection and Prevention" https://coverager.com/insurance-at-the-intersections-of-protection-and-prevention/

325. Susan Wharton Gates, Vanessa Gail Perry & Peter M. Zorn (2002) "Automated underwriting in mortgage lending: Good news for the underserved?" Housing Policy Debate. 13:2, 369-391, DOI: 10.1080/10511482.2002.9521447

326. Bonnie Buchanan, (2019, April 02). "Artificial intelligence in finance." Zenodo. http://doi.org/10.5281/zenodo.2612537

327. "Fukoku Mutual – Insurance firm to replace human workers with AI system" https://www.the-digital-insurer.com/dia/fukoku-mutual-insurance-firm-to-replace-human-workers-with-ai-system-2/

328. Swiss Re Institute's Bohn: "Startups Will Be Sought for Their Automation Prowess "video accessed 28 Jan 2019

329. Randy Bean, Forbes,"Munich Re: How Data and AI Reduce Risk from Global Calamities" Nov 4, 2018. https://www.forbes.com/sites/ciocentral/2018/11/04/munich-re-how-data-and-ai-reduce-risk-from-global-calamities/

330. Cameron Nicol, "Enhancing Private Equity Manager Selection with Deeper Data," CAIA. Issue Q3 2018

331. "The Allure of the Outlier: A Framework for Considering Alternative Investments," Vanguard, 2015

332. "Measuring Institutional Investors' Skill from Their Investments in Private Equity," Cavagnaro, Senoy, Wang and Weisbach, 2016

333. MIT Technology Review, Nanette Byrnes, "As Goldman Embraces Automation, Even the Masters of the Universe Are Threatened" Feb72017 https://www.technologyreview.com/s/603431/as-goldman-embraces-automation-even-the-masters-of-the-universe-are-threatened/?utm_campaign=add_this&utm_source=twitter&utm_medium=post
334. The New York Times, "The Robots are coming for Wall Street" Feb 8, 2016.
335.ING company page, "Katana gives bond traders a cutting edge" 12 December 2017 https://www.ing.com/Newsroom/All-news/Katana-gives-bond-traders-a-cutting-edge.htm
336. Refinitiv company page "Technology in 2024: Is your trading desk ready?" https://www.refinitiv.com/en/products/velocity-analytics-market-data-analysis
337. Daniel Faggella "From Past to Future, Tracing the Evolutionary Path of FinTech – A Conversation with Brad Bailey" https://emerj.com/ai-podcast-interviews/from-past-to-future-tracing-the-evolutionary-path-of-fintech/
338. BAFIN, German Federal Financial Supervisory Authority, "Algorithmic trading and high-frequency trading" 17.07.2019. https://www.bafin.de/EN/Aufsicht/BoersenMaerkte/Hochfrequenzhandel/high_frequency_trading_artikel_en.html
339. Reto Francioni, Robert A. Schwartz. Equity Markets in Transition – The Value Chain, Price Discovery, Regulation, and Beyond. Springer Verlag in Cham, Switzerland 2017. ISBN 9 783319 458465
340. GRAP, the international association of risk professionals. https://www.garp.org/?utm_content=91134049&utm_medium=social&utm_source=linkedin&hss_channel=lcp-872311#!/risk-intelligence/technology/quant-methods/a1Z1W000004wDv5UAEABM
341. Bank of England, "FinTech Proof of Concept MindBridge Analytics Inc. – big data anomaly detection tool" https://www.bankofengland.co.uk/-/media/boe/files/fintech/mindbridge-phase2.pdf?la=en&hash=DB1BB5EC3B4B4900FC205AE00267A3B098CB755A
342. Agrawal, A., Gans, J., & Goldfarb, A. (2018). Economic Policy for Artificial Intelligence (Working paper series (National Bureau of Economic Research: Online); working paper no.24690). Cambridge, Mass.: National Bureau of Economic Research.
343. Alliance for Financial Inclusion (2016). ″Annual Report 2015″ (PDF): 3. Retrieved 20 August 2019.

344. Qatar Financial Center, "Fintech, a global boost for Islamic Finance?" October 5, 2018 https://www.reuters.com/sponsored/article/Fintech-a-global-boost-for-Islamic-Finance
345. Reuters, Fintech a global boost for Islamic Finance brought by Qatar Financial Centre, accessed on 2 May 2019.
346. Rohaya Mat Rahim, Siti & Zuriyati Mohamad, Zam & Abu Bakar, Juliana & Hanim Mohsin, Farhana & Md Isa, Norhayati. (2018). "Artificial Intelligence, Smart Contract and Islamic Finance." Asian Social Science. 14. 145. 10.5539/ass.v14n2p145.
347. IFNFintech, Q1 2019; http://www.ifnfintech.com/secure/admin/uploadedFiles/v3i1-1551941377.pdf accessed 2 May 2019
348. General Council for Islamic Banks and Financial Institutions, "Technology, Innovation and artificial intelligence for the growth of Islamic Finance Industry" http://www.comcec.org/en/wp-content/uploads/2017/10/9-FIN-PRE-7-2.pdf

CHAPTER 9: AI Applications

349. The companies mentioned in this book are not endorsed by the author, and the readers are well advised to perform their own research in order to establish if any of the mentioned companies are suitable to their own specific requirements.
350.MIT Technology Review "AI hits the mainstream" accessed 26 February 2019
351. DataSine company page https://datasine.com accessed on 13 March 2018 - the landing page mentions BNP Paribas
352. O'Reilly, "The next generation of AI assistants in enterprise" August 20, 2018. https://www.oreilly.com/ideas/the-next-generation-of-ai-assistants-in-enterprise
353. Kasito company's report 2018, published with permission
354. JP Morgan company page "Informing Investment Decisions Using Machine Learning and Artificial Intelligence" https://www.jpmorgan.com/global/cib/research/investment-decisions-using-machine-learning-ai
355. disclosure - Clara Durodie is a member of the Advisory Board of StockSmart.
356. The Financial Times, "Fund industry must use the right type of machine learning." 6th February 2019, Marcos Lopez de Prado
357. University of Johannesburg "Artificial intelligence trained to analyze causation " https://phys.org/news/2018-06-artificial-intelligence-causation.html

358."Machine Learning For Stock Trading Strategies" April 14, 2016. https://www.nanalyze.com/2016/04/machine-learning-for-stock-trading-strategies/
359. Idem
360. Sentient "3 Artificial Intelligence Applications from Sentient" March 5, 2016. https://www.nanalyze.com/2016/03/3-artificial-intelligence-applications-from-sentient/
361. Christian Homescu presentation reproduced with permission at https://www.linkedin.com/pulse/ai-finance-summit-cristian-homescu/
362. Elana Margulies-Snyderman, "Trends to watch" 3rd August 2017, accessed 13 March 2019. https://www.eisneramper.com/weinberg-tw-ai-blog-0817/ interview with Michael Weinberg CFA, Chief Investment Officer of MOV37 and Protege Partners
363. Comment on LinkedIn on 15 March 2019 at "Talk to me human" article by Glen D'Amore
364. The New York Times, Arango, T., "I got the news instantaneously, oh boy," 14th September 2008
365. Raghav Bharadwaj "AI for Sentiment Analysis in Finance – Current Applications and Possibilities" February 11, 2019. https://emerj.com/ai-sector-overviews/ai-sentiment-analysis-finance-current-applications-possibilities/
366. Bartholomew, H. (2018) "Symphony bots march on Bloomberg", https://symphony.com/documents/risknet-symphony-bots-march-on-bloomberg.pdf ;
367. CFA "JPMorgan Economists See Promise and Pitfalls in Alternative Data" 16 January 2019. https://www.bloomberg.com/news/articles/2019-01-16/jpmorgan-economists-see-promise-and-pitfalls-in-alternative-data
368. Joshua Thurston "UBS trials Netflix-style investment recommendations" 14 August, 2018. http://citywire.co.uk/wealth-manager/news/ubs-trials-netflix-style-investment-recommendations/a1146444
369. Robin Wigglesworth, The Financial Times, "AllianceBernstein revamps bond trading robot" November 13, 2018. https://www.ft.com/content/d88b01a0-e769-11e8-8a85-04b8afea6ea3
370. Trumid company page https://www.trumid.com
371. Finextra, "Pagaya uses AI to issue $100m in asset-backed securities" 08 February 2019. https://www.finextra.com/newsarticle/33349/pagaya-uses-ai-to-issue-100m-in-asset-backed-securities
372. Pagaya website, 2019

373. Chris Pitt, "Let's use data science to overcome content fatigue" 21 Feb 2019. https://www.thedrum.com/opinion/2019/02/21/lets-use-data-science-overcome-content-fatigue
374. Salesforce's State of Marketing report 2018
375. Forbes, "10 Ways machine learning is revolutionising marketing" https://www.forbes.com/sites/louiscolumbus/2018/02/25/10-ways-machine-learning-is-revolutionizing-marketing/#741b76785bb6
376. IDC white paper, July 2017, https://criteo-2421.docs.contently.com/v/idc-research-can-machines-be-creative accessed 25 February 2019
377. Microsoft, "Azure machine learning customer churn scenario" https://docs.microsoft.com/en-us/azure/machine-learning/studio/azure-ml-customer-churn-scenario accessed 25 February 2019
378. Tibco company page https://www.tibco.com/company accessed 25 February 2019
379. Calibermind company https://www.calibermind.com accessed 26 February 2019
380.Crunchbase, Calibermind company details. https://www.crunchbase.com/organization/calibermind#section-overview accessed 26 February 2019
381. Oracle purchased Datafox in October 2018. https://www.datafox.com accessed 26 February 2019
382. Pega company webpage for Financial Services "Earning trust takes more than a handshake" https://www.pega.com/industries/financial-services/wealth-management accessed 26 February 2019
383. Salesforce company page "Salesforce Einstein Delivers More Than One Billion AI-Powered Predictions Per Day" February 28, 2018. https://www.salesforce.com/company/news-press/stories/2018/2/022818-a/
384. IBM company page "Watson Ads Empower your customers with cognitive advertising to enable intelligent, two-way conversations with your brand" https://www.ibm.com/us-en/marketplace/cognitive-advertising accessed on 26 February 2019
385. Jessica Miller-Merrell "9 Ways to Use Artificial Intelligence in Recruiting and HR" https://workology.com/artificial-intelligence-recruiting-human-resources/
386. Headstart company page https://www.headstart.io/about/
387. In addition to the interview with Ginny Rommetti of IBM, CBNC "IBM AI can predict with 95% accuracy which employees will quit" https://www.cnbc.com/2019/04/03/ibm-ai-can-predict-with-95-percent-accuracy-which-employees-will-quit.html

388. KnowledgeOfficer companies webpage https://app.knowledgeofficer.com
389. EvaAI company webpage http://www.eva.ai
390. Brad Smith, President and Chief Legal Officer, and Harry Shum, Executive Vice President of Microsoft AI and Research Group. The Future Computed, 2018
391. Law 360 " Data Science no longer a luxury for modern law practices" https://www.law360.com/articles/1105413/data-science-no-longer-a-luxury-for-modern-law-practices
392. "AI in legal practices - current applications" 2018 https://emerj.com/ai-sector-overviews/ai-in-law-legal-practice-current-applications/
393. Pew Research, "How artificial intelligence could improve access to legal information" report 2019/01/24
394. Bloomberg, "JPMorgan Software Does in Seconds What Took Lawyers 360,000 Hours" 28 February 2019 https://www.bloomberg.com/news/articles/2017-02-28/jpmorgan-marshals-an-army-of-developers-to-automate-high-finance
395. Luminance company webpage https://www.luminance.com
396. Bank of England "Embracing the promise of fintech" Quarterly bulletin Q2, 2019
397. European Risk Management Council, "Transforming strategic risk management to realise competitive advantage," 2018
398 Liebergen van, Bart, "Machine learning: a revolution in risk management and compliance ?" The CAPCO Institute Journal of Financial Transformation, 2019.
399."Barclays Taps Accelerator Graduate For Risk Simulation" https://www.pymnts.com/news/b2b-payments/2018/barclays-simudyne-risk-simulation
400. Cube company webpage https://www.cube.global
401. Onfido company webpage https://onfido.com/solutions/
402. "UOB and Tookitaki strengthen combat against money laundering through co-created machine learning solution" https://www.uobgroup.com/web-resources/uobgroup/pdf/newsroom/2018/UOB-and-Tookitaki-strengthen-combat-against-money-laundering.pdf
403. MindBridge company press release "Mindbridge AI accelerates its IPO ready strategy with global leadership appointments" https://www.mindbridge.ai/press-releases/mindbridge-ai-accelerates-its-ipo-ready-strategy-with-global-leadership-appointments/
404. Idem
405. Fluidly company website fluidly.com
406. Prophix company website prophix.com

407. Uber engineering, YouTube - Transforming financial forecasting with data science and machine learning, 5 July 2018
408. Goldman Sachs company website https://www.goldmansachs.com/what-we-do/investing-and-lending/principal-strategic-investments/psi-portfolio.html
409. IPSoft company website https://www.ipsoft.com/wp-content/uploads/2016/11/IPcenter_In_Action.pdf
410 https://vimeo.com/54367908

CHAPTER 10: About the Future

411. The more things changes, the more they stay the same (translation from French)
412. The City of London or The City still refers to the financial services industry by the practitioners in London, which was originally located in The City of London, the size of a square mile.
413. The Financial Times "German online bank N26 valued at $3.5bn after $170m investment" July 18, 2019. https://www.ft.com/content/4d31c764-a8c6-11e9-984c-fac8325aaa04
414. Attali, Jacques. A Man of Influence: The Extraordinary Career of S.G. Warburg. New York: Adler & Adler, 1987.
415. Ferguson, Niall. High Financier: The Lives and Time of Siegmund Warburg. New York: Penguin Books, 2010.
416. ESG+T is a concept that I first read about in Andrea Bonime-Blanc's monthly newsletter. Andrea is an internationally recognised risk and cybersecurity expert who explained that as businesses address ESG, T (technology) is increasingly playing a more important role in how ESG is delivered.
417. World Finance, "Sustainable Banks" https://www.worldfinance.com/banking-guide-2018/sustainable-banks
418. https://twitter.com/fchollet/status/1157704165851664384?s=21 published 6:25 pm on 3 Aug 2019
419.Hidden technical debt in machine learning systems, paper submitted at NIPS 2015 Authors Sculley,D. et al.
420. Ocean Protocol company website https://oceanprotocol.com/protocol/
421. Emoshape's founder's online direct presentation at the 2017 Cognitive Finance Group's Annual Executive Retreat

422. idem
423. Li, Rongpeng & Zhifeng, Zhao & Zhou, Xuan & Ding, Guoru & Chen, Yan & Zhongyao, Wang & Zhang, Honggang. (2017). Intelligent 5G: When Cellular Networks Meet Artificial Intelligence. IEEE Wireless Communications. PP. 2-10. 10.1109/MWC. 2017.1600304WC.
424. Patrick Seitz "Artificial Intelligence, 5G Wireless Seen Heralding New Data Age" 1/11/2018. https://www.investors.com/news/technology/artificial-intelligence-5g-wireless-seen-heralding-new-data/
425. Chris Duckett, "Singapore reinstalling technical knowledge to keep vendors honest" https://www.zdnet.com/article/singapore-reinstalling-technical-knowledge-to-keep-vendors-honest/
426. Fintech Magazine "Santander Group puts aside $22.5bn for digital transformation" April 05, 2019. https://www.fintechmagazine.com/banking/santander-group-puts-aside-225bn-digital-transformation
427. Cnet News, Elon Musk Neuralink to Test Human Test Brain Computer Interface in 2020
428. Bergen, Mark 2017, Bloomberg, Google's Quantum Computing Push Opens New Front in Cloud Battle
429. Microsoft book "The Future Computed: Artificial Intelligence and its role in society" https://blogs.microsoft.com/blog/2018/01/17/future-computed-artificial-intelligence-role-society/ accessed 15th September 2019

Appendix: Developments in AI–A Brief History of Achievements

430. Yahoo Finance UK, 16 August 2019, Savers targeted with ads on Google for 'bonds' that put all their money at risk
431. For a very detailed and superbly documented history of AI, I would recommend Nilsson, N. The Quest for Artificial Intelligence: A History of Ideas and Achievements. Cambridge: Cambridge University Press, 2010.
432. Fitch, F. (1944). McCulloch Warren S. and Pitts Walter. "A logical calculus of the ideas immanent in nervous activity. Bulletin of mathematical biophysics." vol. 5 (1943), pp. 115–133. The Journal of Symbolic Logic, 9(02), 49-50.
433. http://www.aiai.ed.ac.uk/events/lighthill1973/

434. "Personal Financial Planning System" https://www.aaai.org/Library/IAAI/1989/iaai89-006.php

435. Bratko, I. (1990). Prolog programming for artificial intelligence (2nd ed., International computer science series). Wokingham: Addison-Wesley. p.vii

436. Buchanan, Bonnie. (2019, April 02). Artificial intelligence in finance, Zenodo

437. Investopedia - Terms : Flash Crash

438. Knight, W. (2018) technologyreview.com The Defense Department has produced the first tools for catching deepfakes, 7th August 2018

INDEX

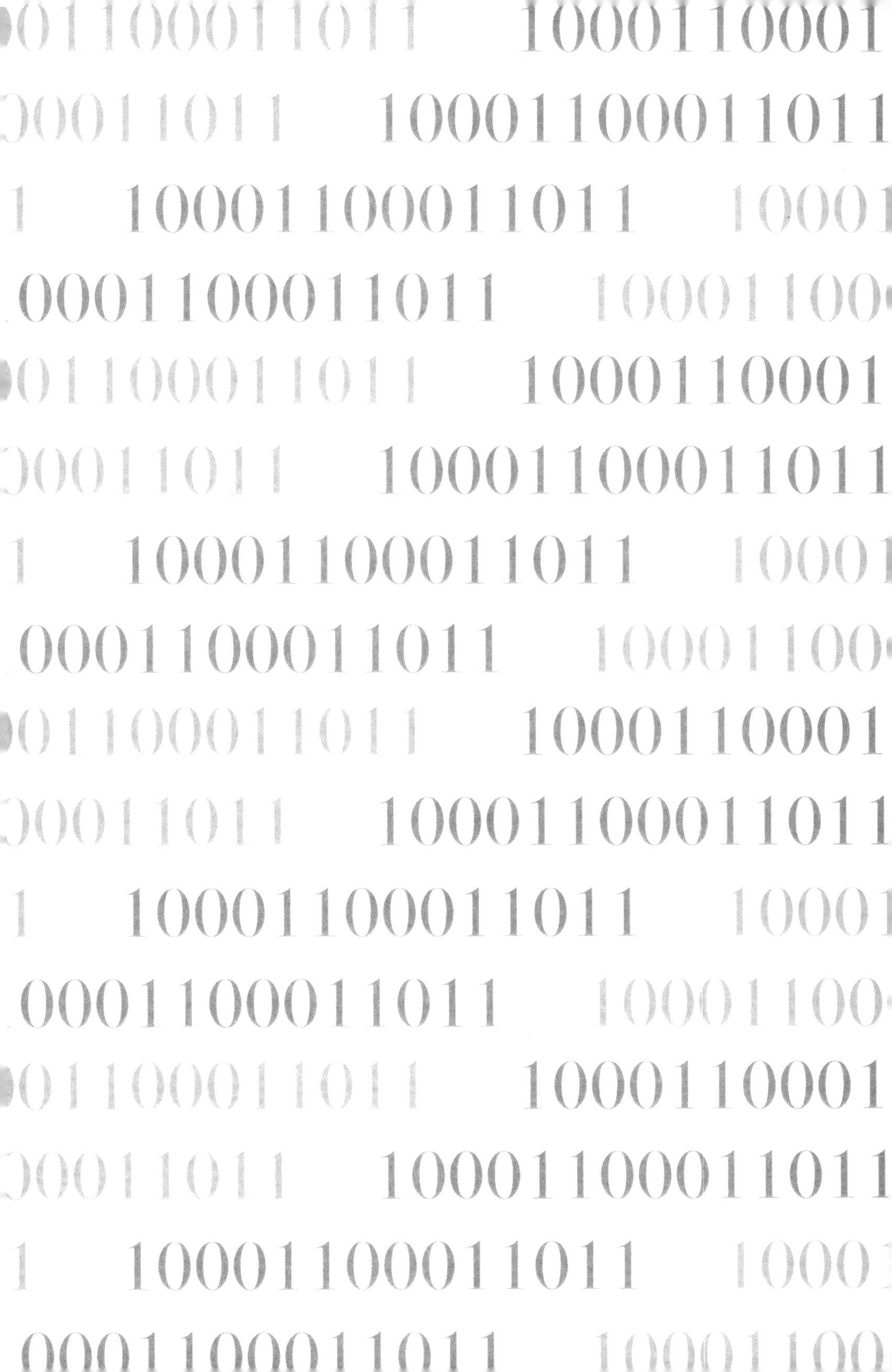

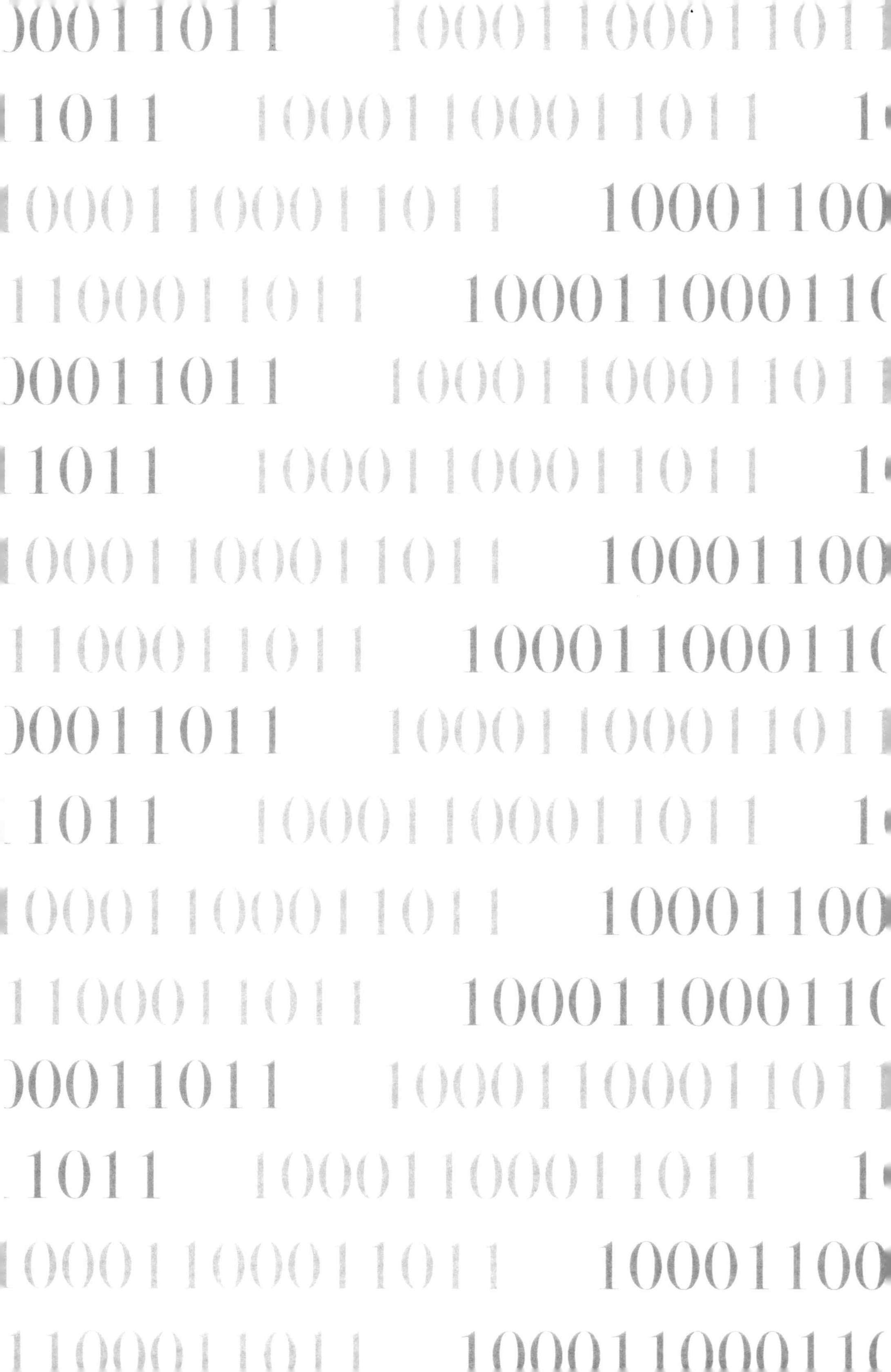